MICHAEL J. REISS and ROGER STRAUGHAN

Improving Nature?

The science and ethics of genetic engineering

CAMBRIDGE
UNIVERSITY PRESS

PUBLISHED BY THE PRESS SYNDICATE OF THE UNIVERSITY OF CAMBRIDGE
The Pitt Building, Trumpington Street, Cambridge CB2 1RP, United Kingdom

CAMBRIDGE UNIVERSITY PRESS
The Edinburgh Building, Cambridge CB2 2RU, UK http://www.cup.cam.ac.uk
40 West 20th Street, New York, NY 10011–4211, USA http://www.cup.org
10 Stamford Road, Oakleigh, Melbourne 3166, Australia

© Cambridge University Press 1996

First published 1996
First paperback edition 1998

Printed in the United Kingdom at the University Press, Cambridge

Typeset in Linotron Ehrhardt 11/13pt [RO]

A catalogue record for this book is available from the British Library

Library of Congress Cataloguing in Publication data
Reiss, Michael J. (Michael Jonathan), 1958–
Improving nature? : the science and ethics of genetic engineering
/ Michael J. Reiss and Roger Straughan.
p. cm.
Includes bibliographical references and index.
ISBN 0 521 45441 7 (hc)
1. Genetic engineering – moral and ethical aspects. I. Straughan,
Roger. II. Title.
TP248.6.R466 1996
174′.9574 – dc20 95-46344 CIP

ISBN 0 521 63754 6 paperback

Little more than a deca__ ___ ___ ____ _____ _____ was hardly known outside research laboratories. By now, though, its use is widespread. Those in favour of genetic engineering – and those against it – tell us that it has the potential to change our lives perhaps more than any other scientific or technological adva_

ethic biolo ing ir _____ ___ __derlying science is explained in a way easily understood by a non-biologist, and the moral and ethical considerations that arise are fully discussed. Throughout, the authors clarify the issues involved so that readers can make up their own minds about these controversial issues.

Michael Reiss is a biologist. He was an undergraduate at Cambridge and then went on to do a PhD and post-doctoral research on evolutionary biology and population genetics. He then trained as a teacher and is currently Senior Lecturer in Biology at Homerton College, Cambridge. He is a Fellow and Vice-President of the Institute of Biology, and is the author of over a dozen books and health education packs. His current research, writing and teaching interests are in science and health education and bioethics. He is also a priest in the Church of England.

Roger Straughan is a moral philosopher. His undergraduate degree also came from Cambridge after which he studied in Bristol and London, where he obtained his PhD. He is currently Reader in Education at the University of Reading. He has written and edited several books on moral education and the philosophy of education and has authored many research articles on the ethics of genetic engineering. He has also acted as ethical advisor to the EU.

Also of interest in popular science

The Thread of Life
SUSAN ALDRIDGE
Remarkable Discoveries!
FRANK ASHALL
Evolving the Mind
GRAHAM CAIRNS-SMITH
Prisons of Light – black holes
KITTY FERGUSON
Extraterrestrial Intelligence
JEAN HEIDMANN
Hard to Swallow: a brief history of food
RICHARD LACEY
An Inventor in the Garden of Eden
ERIC LAITHWAITE
The Clock of Ages
JOHN MEDINA
Beyond Science
JOHN POLKINGHORNE
The Outer Reaches of Life
JOHN POSTGATE
Prometheus Bound:
science in a dynamic steady state
JOHN ZIMAN

To Terence McLaughlin and Raymond Wilson

for their encouragement, support and friendship

Contents

Acknowledgements

We are extremely grateful to the following for valuable discussions, comments on the manuscript or other help: Michael Banner, Irene Beatty, Caroline Berry, Christine Beharrell, Caroline Brown, Ann Bruce, Radha Burnier, John Callaghan, Robert Cook, Jennie Cory, Alan Holland, Ron James, Sheldon Krimsky, Rachel Linfield, Alan Long, Ben Mepham, Colin Miles, Philip Oliver, Iain Palin, Syed Aziz Pasha, the late Keith Pike, Thomas Preston, K. S. Satagopan, Robert Sharpe, Barry Thompson, Michiel Timmerman, Greg Tucker, Henk Verhoog, John Wallwork, Fraser Watts and Sandra Webb.

ILLUSTRATION CREDITS

Figure 1.1. From Perry, G. A. & Hirons, M. J. D. (1970). *Progressive Biology: Book 3*, London: Blandford Press.

Figure 2.7b. Courtesy of Dr A. V. Grimstone.

Figure 5.1. Courtesy New York Times 1993. Reprinted with permission.

Figure 6.1. Modified from Tomkins, S., Reiss, M. J. & Morris, C. (1992). *Biology at Work*. Cambridge: Cambridge University Press.

Figure 6.2. © G. M. G. Gadd. In *Trends in Biotechnology*, 10, 342 (1992) Elsevier Science Publishers Ltd, UK.

Figure 6.3. © C. S. Wheeler.

Figure 7.1. Courtesy of PPL Therapeutics.

Figure 7.2. Photograph by Brian Gunn: IAAPEA.

Figure 7.3. Courtesy IMUTRAN.

Figure 7.4. © C. S. Wheeler.

Figure 7.5. Courtesy of Greenpeace.

Figure 7.6. Reproduced by kind permission of the Pure Food Campaign.

Figure 8.1. Linda Gray from an original by Dean Madden. In Dixon, B. (1993). *Genetics and the Understanding of Life.* © XVIIth International Congress of Genetics.

Figure 8.2. © 1990 Nicole Hollander.

Figure 8.3. © C. S. Wheeler.

1

Introduction

In the early 1980s, little more than a decade ago, the term 'genetic engineering' was hardly known outside research laboratories. By now, though, it is common currency and both those in favour of genetic engineering and those against it tell us that it has the potential to change our lives more than perhaps any other scientific or technological advance.

In this book, we examine the major implications of genetic engineering. We try to explain the underlying science in a way that can be understood by the non-biologist, and we discuss the moral and ethical considerations that arise. Our main hope is to clarify the biological and philosophical issues involved. Occasionally we make recommendations, but mostly we prefer to elucidate the implications of particular views or courses of action to enable you, the reader, to make up your own mind.

In this chapter, we provide a brief introduction both to genetic engineering itself and to the role of moral and ethical considerations. We then outline the contents of later chapters.

What is genetic engineering?

Every organism carries inside itself what are known as genes. This is as true of bacteria and fungi as it is of plants and animals, including ourselves. These genes are codes or messages. They carry information. The information they carry is used to tell the organism what chemicals it needs to make in order to survive, grow and reproduce. Genetic engineering typically involves moving genes from one organism to another. For example, moving a gene from one type of plant to another type. The result of this procedure, if all goes as intended, is that the chemical normally made by the gene in the first type of plant is now made by the second type of plant. Suppose that the gene concerned promotes frost-tolerance, in other words, it reduces the susceptibility of a plant to damage by frost. Then moving it into an important crop plant that is all too often damaged by frosts might have considerable benefits.

Various terms in addition to genetic engineering have been used to describe such activities. These include 'genetic manipulation', 'genetic modification', 'genetic technology', 'recombinant DNA technology' and even 'modern biotechnology'.[1] We have tended to stick with 'genetic engineering'. This is partly because it is the mostly widely used term, and also because, along with genetic modification, it seems to us to be the phrase most exactly to include the two central elements of the new technology, namely the creation (i.e. engineering) of organisms with novel genetic constitutions.

It is easy to see that the potential advantages of genetic engineering are very great indeed. However, the suddenness with which genetic engineering has arrived makes it difficult to assess its validity, and its dangers, as well as its usefulness. For this reason, it helps to examine genetic engineering in the context of traditional biotechnology.

The relationship of genetic engineering to traditional biotechnology

Biotechnology is the application of biology for human purposes.[2] It involves using organisms to provide us with food, clothes, medicines and other products. The phrase 'traditional biotechnology' is used to include all of biotechnology except for those procedures that have only become possible since the mid-1970s or so through advances in genetic engineering and other novel disciplines, such as embryo transfer, molecular biology and tissue culture. Traditional biotechnology is based on activities like the farming of animals and plants and the use of microorganisms in the manufacture of beer, wine, bread, yoghurt and cheese.

Traditional biotechnology has a long history. The domestication of animals and plants seems to have happened independently about 10000 to 8000 BCE in the Middle East, the Orient and the Americas.[3]

Around 10000 to 9000 BCE, the dog was domesticated in Mesopotamia and Canaan. Within a thousand years of this time, goats and sheep were domesticated in Iran and Afghanistan, and emmer wheat and barley were being cultivated in Canaan. Around 8000 to 7000 BCE, potatoes and beans were domesticated in Peru, rice in Indochina and pumpkins in middle America.

By 6000 BCE, the pig and water buffalo had been domesticated in eastern Asia and China, the chicken in southern Asia and cattle in south-eastern Anatolia (modern-day Turkey). At the same time, einkorn wheat was being cultivated in Syria, durum (macaroni) wheat in Anatolia, sugar cane in New Guinea, yams, bananas and coconuts in Indonesia, flax in south-western Asia and maize and peppers in the Tehuacan valley of Mexico. By 6000 BCE, a type of beer was also being made with yeast in Egypt. Indeed, by the second millennium BCE, the Sumerians brewed at least 19 brands of beer – a whole book on the subject survives!

In considering the significance of these and other more recent examples of traditional biotechnology, it helps to appreciate that the following four processes are involved in the farming of domesticated animals or plants:

- Breeding of animals, or sowing of seeds
- Caring for the animals or plants
- Collecting produce (e.g. harvesting, milking, slaughtering)
- Selecting and keeping back some of the produce for the next generation.[4]

For more than 10000 years, therefore, farmers have selected animals and plants. Much of this selection will have been conscious, with farmers often choosing, for example, to breed from larger and healthier individuals. Indeed, genetics is probably a much older science that is generally realised. Figure 1.1 shows a sketch of a clay tablet dating from about 3000 BCE. The tablet is

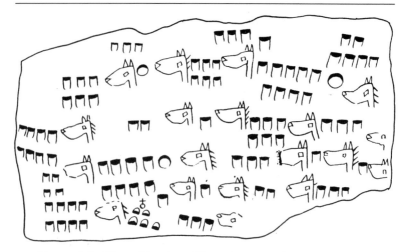

Figure 1.1. Clay tablet from Elam showing what are probably breeding records of domesticated donkeys. Notice the various types of mane. The script is known as proto-Elamite and has not yet been deciphered.

from Elam (now in south-eastern Iran) and appears to show a breeding record of domesticated donkeys.[5] However, much of the selection by farmers will have been unconscious, as farmers unwittingly chose, for example, animals that were tractable or tolerant of overcrowding.

The fact that traditional biotechnology has such a long history might lead one to conclude that perhaps too much fuss is being made about genetic engineering. After all, traditional biotechnology often involves the transfer of genes in a way that would not happen in nature. That happens every time a farmer selects a bull to mate with cows, and every time a plant breeder dusts the pollen from one plant onto the female sex organ of another plant. Indeed, such traditional selective breeding has achieved dramatic results. Think of all the different breeds of dogs. Toy poodles, bulldogs and the St Bernard all stem from the same ancestors.

Traditional biotechnology has changed certain plants very greatly too. The modern wheat used in bread making is so different from native wheats that scientists are still uncertain as to its precise ancestry. What is clear, though, is that it results from at least two interspecific crosses. In other words, on at least two separate occasions, thousands of years ago, people succeeded in breeding one species of wheat with another species. The net result is that today's bread wheat contains approximately three times the number of genes as wild wheats found in the Middle East.

However, although traditional biotechnology can result in major alterations in the genetic make-up of organisms, it differs from genetic engineering in at least three important respects.

First, although traditional biotechnology sometimes involves crossing one species with another, these species are always closely related. To the non-expert, the plant species crossed to make modern bread wheat all look much the same. Indeed, botanists classify them as being very closely related. This is markedly different from genetic engineering where, already, to give just two examples, human genes have been put into pigs and genes from bacteria into plants.

Secondly, the pace of change in traditional biotechnology is much slower than in genetic engineering. We are already at the point where a gene from one organism can permanently be inserted into the genetic material of another organism within a period of weeks. Traditional biotechnology, by comparison, works on a time scale of years.

Thirdly, genetic change as a result of traditional biotechnology happened to only a relatively small number of species, namely those that provide us with food and drink, such as crop plants, farm animals and yeasts. Genetic engineering is far more ambitious. It seeks to change not only the species that provide us with food and drink but also those involved in sewage disposal, pollution control and drug production. It also seeks to create microorganisms, plants and animals that can make human products, such as insulin, and even, possibly, to change the genetic make-up of humans.

Moral and ethical considerations

Having outlined, very briefly, the nature of genetic engineering and its relationship to traditional biotechnology, one further question in this introductory chapter needs to be tackled: why is it necessary to investigate *moral* and *ethical* concerns about genetic

engineering? Some people argue that the key questions and problems here are scientific and commercial ones, best left to those with
expertise in science and commerce. Ethical debate may provide a
stimulating pastime for moral philosophers, but does it have any
practical importance in the real world?

Two responses can be made to such queries. First, no new
scientific or technological development can claim immunity from
ethical scrutiny. The fact that new technologies exist does not
mean that they necessarily *ought* to be employed. The pursuit of
new knowledge and techniques can never be given a total ethical
carte blanche, as a hypothetical example can readily demonstrate:
'No matter how interested a researcher may be in investigating the
effects of massive doses of bomb-grade plutonium on preschool
children, it is hard to imagine that anyone thinks he should be
allowed to do so'.[6]

This of course raises difficult questions about restrictions on
scientific research, which we will explore more deeply later. At this
point, however, we only need to note that science cannot be
pursued in a moral and ethical vacuum in any society that claims to
be healthy or civilised; the universal condemnation of so-called
'medical research' as pursued in various countries, including Nazi
Germany, during times of war supports this view.

Moreover, it is not the case that scientists, technologists and
industrialists generally *want* to operate in such a vacuum. As
rational human beings, the great majority of such people appreciate the moral dimension to life as well as anyone else does. Moral
and ethical questions, then, cannot be by-passed or ignored in
genetic engineering, any more than they can elsewhere.

The second response to the query as to whether ethical debate
has any practical importance in the real world is to acknowledge
that moral and ethical concerns are of considerable practical
importance in informing and influencing public attitudes towards
modern biotechnology. Our attitudes are not, of course, wholly
determined by our moral views alone, for other more pragmatic or
self-interested factors will also play a part. Nevertheless, our moral
views about any subject will clearly exert a powerful effect upon our
perceptions of that subject and upon our attitudes towards it, for
our moral beliefs play an integral part in how we see and interpret

the world, which in turn helps to shape our choices and behaviour.

However, there remains the question: what is so special about genetic engineering? In other words, why a book on the ethical concerns of this particular subject? One possible answer is to suggest that perhaps there is nothing *that* special about genetic engineering. Rather, every significant scientific advance throws up attendant problems and possibilities that need ethical reflection. This is as true of genetic engineering as it is of nuclear power, *in vitro* fertilisation, the genetics of race, pollution, conservation and the many other subjects of public concern that science has brought into prominence.

However, it can also be argued that perhaps there *is* something distinctive about genetic engineering. For one thing, its scope seems near endless. We are promised that genetic engineering will revolutionise agriculture, medicine, the food industry and much else besides. For another, we are told by some that its potential for harm is immense. Then there is the pace of change: progress in genetic engineering seems to be happening so fast that perhaps ethical reflection is needed now before it is too late. Indeed, some claim that our scientific understanding has already outstripped our powers of moral comprehension. Finally, many aspects of genetic engineering strike at the very heart of our lives. Genetic engineering raises issues about the nature of life itself, about what it is to be human, about the future of the human race and about our rights to knowledge and privacy.

For all these reasons, writings on the ethical implications of genetic engineering are needed, and needed now.

Arrangement of the book

We have divided this book into three sections. Part 1 includes Chapters 2, 3 and 4. Chapter 2 provides an overview of the practicalities of genetic engineering; it covers the essential biology of the technology. Chapter 3 discusses moral and ethical concerns and considers how they might be distinguished, categorised and evaluated. Chapter 4 tackles theological concerns. There will be some readers for whom a treatment of theological issues is seen as an irrelevancy; equally, there will be other readers for whom theological considerations are paramount. Our aim in this chapter is principally to discuss the variety of theological approaches to the issue of genetic engineering. We hope to assess the validity of the arguments that have been put forward and to show how different conclusions have been reached by different theological commentators, depending on their suppositions.

Part 2 is divided into four chapters: on the genetic engineering of microorganisms, plants, non-human animals and, the last, humans. Within each of these chapters, we look at the range of applications and the possible implications of genetic engineering and then focus on a few case studies in more depth. These case studies are described and then analysed in terms of the ethical and theological principles outlined in Chapters 3 and 4.

Part 3 contains just one chapter, Chapter 9. In it we look at the importance of education. We examine whether more education about genetic engineering is necessary and clarify precisely what is meant by education, and how it differs from the mere provision of information. We end with notes, which include references, and an index.

Our aim in this book is to provide a balanced overview of the practice and potential of genetic engineering, exploring the scientific and philosophical principles involved. Our intention is to address the question posed in our title: does genetic engineering improve on nature? We hope that our account is readable but does not oversimplify. We would be delighted if our efforts help to demystify the subject and to clarify the current issues and possible ramifications. It is our belief that genetic engineering is here to stay, but that its scope, and our response to it, have yet to be decided.

PART 1

2

The practicalities of genetic engineering

Genetic engineering involves the direct, intentional alteration of the genetic material of organisms. Understanding the ethical implications of genetic engineering requires a certain amount of knowledge about the practicalities of genetic engineering: how it is carried out, what the benefits are and what the difficulties and attendant dangers are. But before we can say very much specifically about genetic engineering, we need to appreciate something about the nature of the genetic material and how it works.[1]

The nature of the genetic material

Multicellular organisms, such as ourselves and all the other creatures that are visible to the naked eye, contain large numbers of cells (Figure 2.1*a,b*). An adult human has about ten thousand thousand million of these cells, each one invisible unless viewed under a microscope. Cells get their names from their superficial resemblance to the cells of medieval enclosed monks, as noted by the seventeenth century scientist Robert Hooke. This resemblance is because biological cells are bounded. All cells are enclosed within a membrane, which helps regulate the passage of substances into and out of the cell. Plant cells also possess a cell wall outside their membrane. A plant's cell wall is far thicker than its membrane and provides mechanical support and protection.

Near the centre of most cells is located the nucleus, a structure

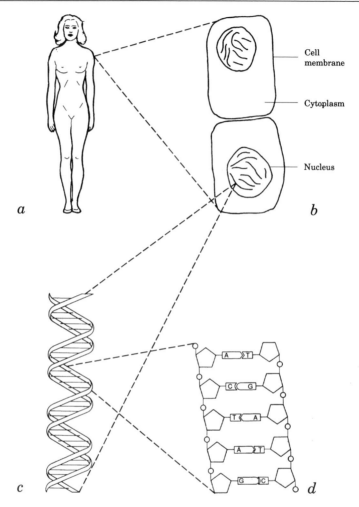

Cell
membrane

Cytoplasm

Nucleus

a

b

c

d

Figure 2.1. (*a*) Organisms are made up of cells. (*b*) A cell is surrounded by a membrane and usually contains a single nucleus. The diagram shows a generalised animal cell, such as is found in the layer of cells that gives rise to skin. (*c*) Chromosomes are found in the nucleus. Chromosomes contain nucleic acid (DNA) and protein. For simplicity the protein has been omitted from this diagram. (*d*) The structure of DNA in more detail. There is a backbone consisting of alternating sugars and phosphate groups. To this backbone bases are attached, represented here by the letters A, C, G and T.

that helps control the activities of the cell in much the same way that our brain helps control the activities of our body. The rest of a cell, apart from the nucleus, is called the cytoplasm, but it is in the nucleus that the chromosomes are found (Figure 2.1*c*). Chromosomes are thin strands containing both nucleic acid and protein. By the end of the nineteenth century, biologists had observed and described chromosomes under the light microscope. However, it was only in 1902 that Walter Sutton in the USA and Theodor Boveri in Germany suggested that chromosomes might carry genetic information. Now we know that the chromosomes do indeed carry the genetic information, though it wasn't until 1944 that Oswald Avery showed that this information is carried by the nucleic acids of the chromosomes, rather than by the proteins.

So by 1944, it was known that the genetic information inherited by offspring from their parents was carried by the nucleic acid component of the chromosomes situated in the nuclei of cells. The specific cells involved are those that give rise to the gametes from which a new offspring results. In humans, the relevant cells are those in the ovary that give rise to eggs and those in the testes that give rise to sperm. However, it wasn't until 1953 that the structure of the nucleic acid component of chromosomes was discovered. It was this discovery that gave birth to the field of molecular genetics and subsequently led to the advent of genetic engineering.

The structure of DNA

Several different types of nucleic acid are known. The type found in chromosomes is called deoxyribonucleic acid, generally abbreviated to DNA. In 1953, work by Francis Crick, Rosalind Franklin, James Watson and Maurice Wilkins lead to the determination, by James Watson and Francis Crick, of the structure of DNA. Working with data obtained by Franklin and Wilkins, Watson and Crick realised that DNA was a double helix (Figure 2.1*d*). DNA consists of two long strands twisted around each other. Humans have 46 chromosomes in the great majority of their cells, which means that each cell has 46 long DNA molecules.

The structure of DNA is really quite simple. Each of the two strands consists of a backbone composed of alternating sugars and phosphate groups. The details of this backbone need not concern us greatly as its structure is identical in practically every organism on this planet. What are of more interest are the bases, as they are known, represented in Figure 2.1*d* by the letters A, C, G and T. It is the arrangement of these bases that determines the characteristics of the organism.

The letters A, C, G and T, stand, respectively, for adenine, cytosine, guanine and thymine. The chemical structure of these four bases is given in Figure 2.2. Again, the details are not crucial for our purposes, but you can see that each one is made up solely of carbon (C), hydrogen (H), nitrogen (N) and oxygen (O) atoms. What is important is that adenine and thymine bases readily form a weak chemical link, or bond, with each other, as do the cytosine

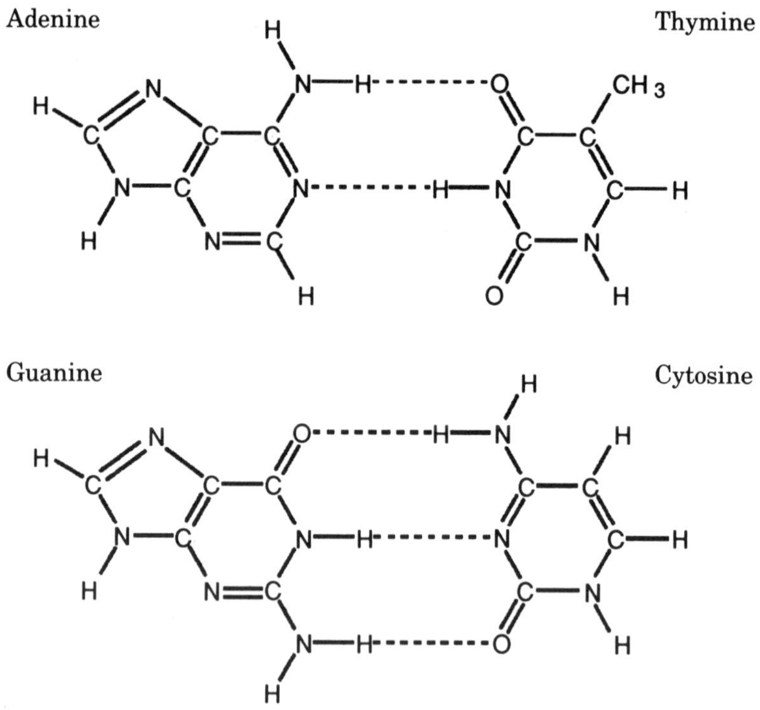

Figure 2.2. The structural formulae of the four bases in DNA. Note that adenine pairs with thymine, and cytosine with guanine.

and guanine bases. This means that an adenine on one strand of a DNA molecule always bonds with a thymine on the other strand, a thymine with an adenine, a cytosine with a guanine and a guanine with a cytosine. This bonding is often referred to as base pairing, as the bases bond in pairs, A with T and C with G.

DNA is a code

DNA is a message or code. The arrangement of the four bases determines the chemicals made by a cell in a remarkable way. An analogy may help. You can read this book because English has an alphabet of 26 letters. The arrangement of these 26 letters, together with some punctuation, allows us, the authors, to communicate with you, the reader. In much the same way, DNA is written in an alphabet, but an alphabet of only four letters. What is more, all the words are three letters long! The details of how these three-letter 'words' determine the chemicals made by a cell are quite subtle but well worth understanding, both for their intrinsic interest and because understanding this makes it possible to see how genetic engineering works.

From what we have said so far, you will remember that DNA is found in the chromosomes of the cell, located in the nucleus. DNA is responsible for the structure of the proteins made by the cell. Proteins are a very diverse group of biological compounds. Examples include haemoglobin, the protein that carries oxygen in our red blood cells, antibodies, the proteins that help protect us against invading microorganisms, and actin and myosin, proteins found in muscle. Not only that, but practically all the enzymes that control our chemical reactions are proteins. For example, salivary amylase, the enzyme in the mouth that helps begin the digestion of our food, is a protein. Indeed, enzymes are responsible for the synthesis and breakdown of practically all the chemicals in a cell. This means that proteins are the key to the functioning of a cell. Cells make thousands of different proteins, each with its own specific function. It is the variability of the structure of DNA that enables a cell to make all these different proteins.

Protein synthesis

Proteins are made by DNA via intermediaries known as messenger RNA and transfer RNA. Messenger RNA and transfer RNA are both examples of ribonucleic acids. Ribonucleic acids (RNAs) are nucleic acids like DNA but have a different sugar in their backbone (ribose, rather than the deoxyribose of DNA). Messenger RNA is so called because it acts as a messenger. It is made in the nucleus and then moves out of the nucleus into the cytoplasm to structures called ribosomes. Here, on the ribosomes, the information carried by the messenger RNA is used to determine the structure of proteins.

What happens is that, in the nucleus, a portion of the DNA double helix unwinds and the bonds between the pairs of bases break. This separates and exposes the two strands of DNA. The information on one of these strands is generally inactive, being needed chiefly when the cell divides, though it also serves as a back-up in case the other strand is damaged. However, the other active strand allows the manufacture of messenger RNA. With the DNA double helix unwound and temporarily separated in this region, the exposed bases allow the synthesis of a piece of messenger RNA, as shown in Figure 2.3. You can see that the order of the bases on the DNA strand determines the order of the bases on the messenger RNA strand. The name given to the process in which DNA makes messenger RNA is *transcription*.

Messenger RNA has, like DNA, a sugar-phosphate backbone with projecting bases. Three of the bases found in messenger RNA are the same as those found in DNA: adenine, cytosine and guanine. However, instead of thymine, messenger RNA has a different base: uracil (abbreviated U).

We are close now to the end of the story of how DNA makes proteins. Having moved from the nucleus to a ribosome in the cytoplasm of a cell, the messenger RNA now waits for the transfer RNAs to play their part. Transfer RNAs are so called because they carry (i.e. transfer) amino acids from other parts of the cytoplasm to ribosome–messenger RNA associations. The significance of this is that amino acids are the building blocks of proteins. You can see in Figure 2.4 that base pairing again plays a crucial role. The

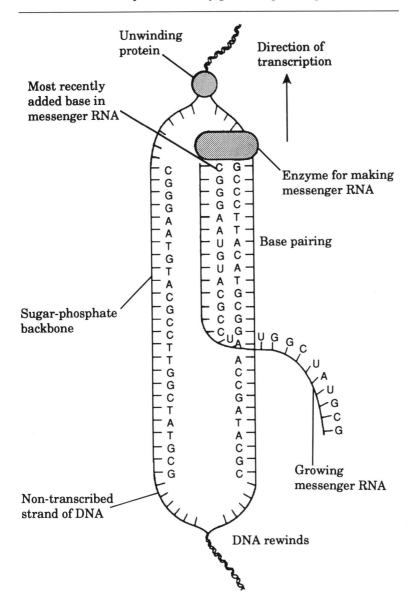

Figure 2.3. The synthesis of messenger RNA by DNA, known as transcription.

1. Messenger RNA synthesised alongside DNA strand in nucleus

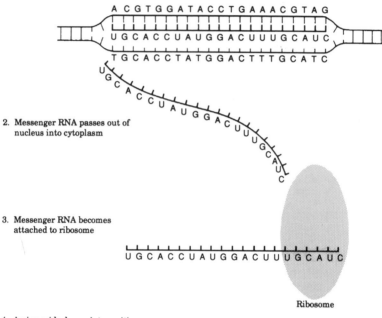

2. Messenger RNA passes out of
   nucleus into cytoplasm

3. Messenger RNA becomes
   attached to ribosome

4. Amino acids drawn into position
   by transfer RNA molecules

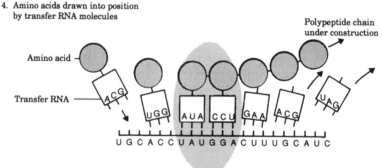

Figure 2.4. Diagram summarising how DNA makes protein via
messenger RNA.

transfer RNAs base pair with different portions of the messenger
RNA. This brings the amino acids that they carry into alignment.
The amino acids are then joined together, and a chain of amino
acids results. A chain of amino acids constitutes a polypeptide

chain. Some polypeptide chains are themselves proteins. Often, though, a protein consists of two or more different polypeptide chains associated together, sometimes with another smaller chemical attached. The name given to the process in which messenger RNA makes a polypeptide chain is *translation*.

The details of the code

We said earlier that the DNA code is written in an alphabet of four letters in words of three letters. We have seen that the four different letters in DNA are A, C, G and T. These, respectively, correspond, in messenger RNA, to the four letters: U, G, C and A. As in any human language, the words are read only in one direction (e.g. left to right in English, right to left in Arabic). As there are four possible letters in the first letter of a DNA word, four in the second and four in the third, there is a total of $4 \times 4 \times 4 = 64$ different possible words. Each three-letter word on a messenger RNA chain is known as a codon.

Table 2.1 presents what is known as the genetic dictionary. This shows the relationship between the 64 possible codons of messenger RNA and the amino acids that result. In all, only 20 amino acids can be made in this way. Other, rarer, amino acids are made by cells from these 20 amino acids. You can see from Table 2.1 that most of the amino acids are coded for by more than one codon. Note too that three of the codons do not actually code for an amino acid. Instead they cause a polypeptide chain to stop getting longer at that point. These three codons are analogous, therefore, to the full stops with which we punctuate our writing. Not surprisingly, they are referred to as stop codons.

It is difficult to generalise, but a typical polypeptide chain might be 200 amino acids in length. This requires a messenger RNA of 603 bases (three bases for each amino acid plus a stop codon). The piece of DNA that corresponds to this messenger RNA is called a gene. Different organisms have different numbers of genes. Bacteria have only a thousand or so, whereas multicellular animals and plants have tens of thousands.

To sum up, we have seen how the structure of DNA allows different messenger RNAs to be produced. These in turn enable

Table 2.1 *The genetic dictionary, showing the relationship between*
messenger RNA triplets (codons) and the amino acids for which they code

First base	Second base				Third base
	U	C	A	G	
U	UUU ⎱ Phe UUC ⎰	UCU ⎱ UCC ⎰ Ser	UAU ⎱ Tyr UAC ⎰	UGU ⎱ Cys UGC ⎰	U C
	UUA ⎱ Leu UUG ⎰	UCA ⎱ UCG ⎰	UAA Stop UAG Stop	UGA Stop UGG Trp	A G
C	CUU ⎱ CUC ⎰ Leu	CCU ⎱ CCC ⎰ Pro	CAU ⎱ His CAC ⎰	CGU ⎱ CGC ⎰ Arg	U C
	CUA ⎱ CUG ⎰	CCA ⎱ CCG ⎰	CAA ⎱ Gln CAG ⎰	CGA ⎱ CGG ⎰	A G
A	AUU ⎱ Ile AUC ⎰	ACU ⎱ ACC ⎰ Thr	AAU ⎱ Asn AAC ⎰	AGU ⎱ Ser AGC ⎰	U C
	AUA ⎰ AUG Met	ACA ⎱ ACG ⎰	AAA ⎱ Lys AAG ⎰	AGA ⎱ Arg AGG ⎰	A G
G	GUU ⎱ GUC ⎰ Val	GCU ⎱ GCC ⎰ Ala	GAU ⎱ Asp GAC ⎰	GGU ⎱ GGC ⎰ Gly	U C
	GUA ⎱ GUG ⎰	GCA ⎱ GCG ⎰	GAA ⎱ Glu GAG ⎰	GGA ⎱ GGG ⎰	A G

Note: Ala, alanine; Arg, arginine; Asn, asparagine; Asp, aspartic acid;
Cys, cysteine; Gln, glutamine; Glu, glutamic acid; Gly, glycine; His, histidine;
Ile, isoleucine; Leu, leucine; Lys, lysine; Met, methionine; Phe, phenylalanine;
Pro, proline; Ser, serine; Thr, threonine; Try, tryptophan; Tyr, tyrosine;
Val, valine.

different polypeptide chains, and hence proteins, to be synthesised. Many of these proteins are enzymes. These enzymes control chemical reactions, causing other compounds, such as carbohydrates, fats and even nucleic acids themselves, to be synthesised, degraded or altered, as circumstances require. In short, the essence of the cell's structure and functioning is determined by the order of the four bases on its DNA.

Mutations

The great majority of biologists believes that over a period of several thousand million years all the different organisms that we see today have evolved from the simple chemicals found on the Earth all that time ago. In particular, biologists believe that all today's organisms once shared a common ancestor. This would explain, for instance, why the same genetic dictionary holds for bacteria, for fungi, for plants and for animals. A mutation is a change in the genetic material of an organism. It is generally thought that mutations are the starting point for evolution. In the great majority of cases, mutations are harmful. However, it is believed that, on rare occasions, a mutation confers a selective advantage on the organism possessing it. Over time, such a mutation may become established, replacing the earlier form of the genetic material.

Mutations can be of various kinds. The simplest is where a single base is replaced. It might be thought that such a tiny change could have no functional consequences, but we know this is not necessarily the case. Despite the fact that each human cell has some three thousand million base pairs in its DNA, the replacement of just one particular thymine base by adenine can lead to the serious, often fatal, condition of sickle cell anaemia. In sickle cell anaemia, instead of the sixth amino acid of each β-haemoglobin chain found in our red blood cells being the usual glutamic acid, it is valine. This is because the triplet of DNA bases on the relevant gene is CAT instead of CTT. CAT leads to the messenger RNA sequence GUA instead of the normal GAA. As you can see from Table 2.1, the result is a valine instead of a glutamic acid at this point.

This type of mutation is called a substitution, as one base is substituted for another. Such a substitution is an example of a point mutation, which is the name given to a mutation involving a change in only a single base. Other instances of point mutations are deletions, when a single base is removed, and insertions, when a single base is added. Deletions and insertions generally have more serious consequences than substitutions. This is because a substitution, unless it results in a stop codon or changes a stop codon into a codon for an amino acid, only results, at worst, in the replacement of a single amino acid in the polypeptide chain by another. However, deletions and insertions have knock-on effects. Because the genetic code is always read in words of three letters, the effect of the insertion or deletion of a base is to alter most of the subsequent amino acids in the polypeptide chain.

The types of mutation we have discussed so far are collectively referred to as gene mutations, as they occur within a gene. Larger changes to the genetic material are known as chromosome mutations. For example, a whole chromosome may be lost, or an extra one found. As we mentioned earlier, humans typically have 46 chromosomes in each of the cells of their body. Occasionally mistakes prior to conception result in an embryo with 45 or 47 per cell. In the great majority of cases, such a chromosome complement is incompatible with life and a natural miscarriage results early in the pregnancy. However, an extra copy of one of the smallest of the human chromosomes, chromosome 21, is compatible with life. Possession of an extra copy of chromosome 21 results in Down's syndrome.

Plants are less sensitive than animals to the consequences of chromosome mutations, and even in animals loss of part of a chromosome, or the possession of an extra copy of part of a chromosome (known as a duplication) is often compatible with life. A further example of a chromosome mutation is when part of a chromosome is inverted, i.e. turns through 180°. Suppose that we label the genes on a particular chromosome A, B, C, D, E, F, G, H, I, J, K, L, M, N, O, P, Q, R. Then a chromosome inversion, as it is known, might result in the sequence: A, B, N, M, L, K, J, I, H, G, F, E, D, C, O, P, Q, R. Generally speaking, such inversions have less serious implications than deletions or duplications.

DNA replication and cell division

So far we have dealt with the structure of DNA, considered how DNA makes proteins and looked at the occurrence of mutations. DNA has one other important function in addition to protein synthesis, and that is to synthesise, or replicate, itself. Consider a human fertilised cell just after the moment of conception. Such a cell contains 46 chromosomes, 23 of them from the mother, 23 from the father. If all goes well, such a cell will result, some nine months later, in the birth of a healthy baby.

It is obvious that during these nine months, a tremendous number of cell divisions have to take place. The original cell divides in two, giving rise to two daughter cells, as they are known. These in turn divide in two giving rise to a total of four cells, and so on. The name given to each of these cell divisions is mitosis. Mitosis has one function only: to produce daughter cells that are identical to the cell that gave rise to them. Each daughter cell has the same number of chromosomes and the same DNA complement as the cell from which it arose.

Watson and Crick realised that the structure of DNA ideally suited it for replication. The process is illustrated in Figure 2.5. As in protein synthesis, the DNA double helix unwinds. This time, though, instead of messenger RNA being synthesised on one of the two strands, each of the two strands synthesises a complementary DNA strand. The net result is the formation of two identical DNA double helices from the one DNA double helix. All the DNA in a cell replicates in this way, thus ensuring that each daughter cell has exactly the same DNA as the cell that gave rise to it.

There is a second type of cell division in addition to mitosis. This type of cell division occurs in the production of gametes and is known as meiosis. Meiosis actually consists of two successive cell divisions, so that a single parent cell gives rise to four daughter cells. However, although meiosis consists of two cell divisions, DNA replication takes place only once. The result is that each daughter cell contains only half the DNA of the original parent cell. In humans this means that each egg and each sperm contains only 23 chromosomes, instead of the usual 46. Then, at fertilisation, a sperm and an egg fuse and the single cell

Improving Nature?

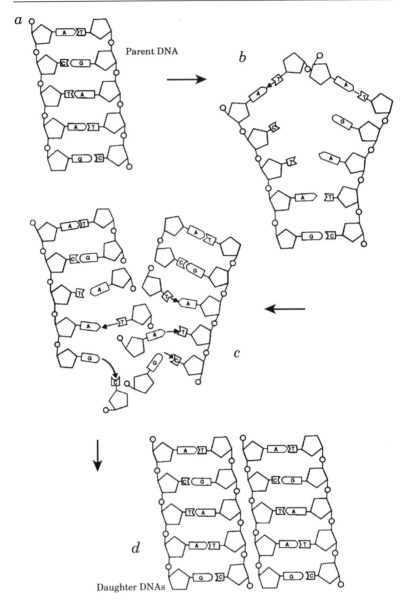

Figure 2.5. Replication of DNA. Now how the process relies on base pairing and results, providing there are no errors, in the production of two identical daughter DNA molecules.

that results has its chromosome complement restored to the usual 46.

So, aside from the gametes, each human cell contains 23 pairs of chromosomes, one member of each pair coming from the mother and one from the father. Consider, for example, chromosome 1, as it is called; this is the longest of the human chromosomes. Each human cell, apart from the gametes, contains one maternal copy of chromosome 1 and one paternal copy. The structure of the DNA of these two DNA double helices will be very similar, but there will be differences resulting from mutations over the course of time.

Imagine that a particular region of chromosome 1 is the gene responsible for the synthesis of a certain polypeptide chain. Suppose now that in a particular individual human the maternal copy of this chromosome is normal but the paternal copy is faulty. There are, therefore, two different forms of the gene: one normal, one faulty. Different forms of genes are called alleles. In this case, the maternal allele is normal, the paternal allele faulty. In such a situation, what one often finds is that the individual is perfectly healthy even though one of the two copies of the gene they have in every one of their cells is faulty. This, for instance, is the case with both sickle cell anaemia and cystic fibrosis. What happens is that the normal allele in each cell produces enough of the required polypeptide to compensate for the faulty allele. In such cases, individuals who are perfectly healthy despite having in each of their cells one faulty copy of the gene are said to be carriers. At the same time, the normal form of the gene is said to be dominant, while the faulty allele is said to be recessive.

The problems with faulty alleles only really arise when both copies of an individual's genes are faulty. So, for example, people with sickle cell anaemia have the single base substitution we discussed earlier (p. 21) on both the maternal and the paternal copies of the gene responsible for making the β-haemoglobin chain. Similarly, people with cystic fibrosis have a faulty gene inherited both from their mother and from their father.

Although most gene mutations, such as those responsible for sickle cell anaemia and cystic fibrosis, are recessive, some are dominant. An example is the gene mutation responsible for

Huntington's disease. In such cases, a person is affected even if each of their cells still has a copy of the normal allele; simply possessing a single copy of the faulty allele is enough to cause the condition.

Differences between organisms in how the genetic material operates

Although the fundamental principles of protein synthesis and DNA replication are the same in all organisms, there are certain differences that have significant implications for the genetic engineer. These differences result from differences in the structure and packaging of genetic material in bacteria, viruses and other organisms, to which we shall now turn.

Bacteria

Bacteria, which are always unicellular, have much in common with other organisms, whether unicellular or multicellular. However, bacteria are much smaller than even the smallest unicellular animals and plants. Figure 2.6 shows a generalised bacterium. Comparing this figure with Figure 2.1, the following distinctive features of bacteria can be noted:

- Bacteria lack a nucleus
- Bacteria only have the one chromosome and this is in most cases circular
- Bacteria have small circular extra-chromosomal bits of DNA, called plasmids.

There are other difference too, but as these mainly relate to features that other organisms possess and bacteria lack, they will be discussed below under the heading *Other organisms*.

As far as genetic engineering goes, the most significant feature of bacteria is that they carry plasmids. These small circular bits of DNA have a variety of functions. Often they carry genes that code for antibiotic-destroying enzymes. The possession of plasmids, therefore, helps disease-causing bacteria to evolve resistance to

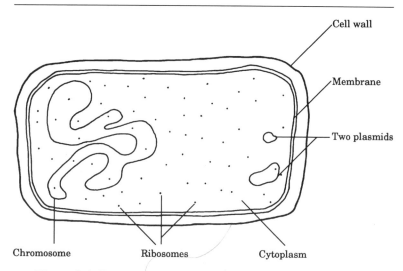

Figure 2.6. Structure of a generalised bacterium. In addition to differences in structure, bacteria are much smaller than animal and plant cells.

antibiotics. Plasmids also play a role in the transfer of genetic information between bacteria. Bacteria have various ways of exchanging genetic material. Many bacteria can simply pick up bits of DNA from their environment and insert it into their plasmids. Further, this DNA doesn't have to come from a bacterium belonging to the same species. Indeed, it doesn't have to come from a bacterium: many bacteria are quite adept at picking up DNA, whatever the source.

Viruses

Viruses are remarkable 'things'. Many biologists don't even like to refer to them as organisms, on the grounds that they are so simple that they cannot exist independently of other organisms. All viruses are parasites. Many of them can survive away from other organisms for long periods of time, but in order to reproduce they must infect other organisms.

There is an almost bewildering variety of types of virus, though all are very small. One example is shown in Figure 2.7 and is known as a bacteriophage as it 'eats' bacteria. To all intents and purposes

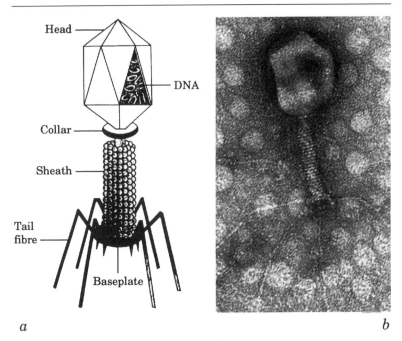

Figure 2.7. (*a*) Diagram of a typical bacteriophage – a virus that attacks bacteria. (*b*) Photograph of a bacteriophage taken under the electron microscope. Viruses are even smaller than bacteria.

this virus, like all others, is little more than a package of genetic material surrounded by a structure made by the virus. This structure serves two functions: protection and a means of infection. The life cycle of this virus is as follows:

- The virus attaches to the surface of a bacterium
- The sheath of the virus contracts and its DNA is injected into the hapless bacterium
- Enzymes produced by the virus destroy the bacterium's DNA
- The virus takes over the running of the bacterium, causing it to produce all the viral proteins coded for by the virus' DNA and many copies of the DNA itself
- Dozens, sometimes hundreds, of new viruses assemble inside the bacterium

- An enzyme produced by the virus causes the bacterium to burst, releasing the new viruses.

There is one further possibility in the life cycle of a bacteriophage. Instead of immediately taking over the machinery of the bacterial cell, the virus may insert its DNA into the DNA of the host bacterium. During cell reproduction, the host cell faithfully replicates the viral DNA along with its own. Then, at some point in the future, the viral DNA may excise itself from the host DNA, take over the cell and start to produce many new viruses.

Bacteriophage viruses have DNA as their genetic material, just as we do. However, some viruses have RNA as their genetic material. An example is HIV, human immunodeficiency virus, the virus that can lead to AIDS. When HIV attacks a susceptible human cell, the following sequence of events occurs:

- HIV attaches to the surface of the cell
- Viral RNA enters the cell along with an enzyme called reverse transcriptase
- Reverse transcriptase enables viral RNA to make complementary viral DNA
- Viral DNA inserts into host's DNA
- At some future point, which may be years later, viral DNA makes complementary viral RNA
- This viral RNA acts as messenger RNA and makes the viral proteins
- Viral proteins assemble around viral RNA, resulting in multiple HIVs
- HIVs break out of cell.

Viruses, such as HIV, that have RNA as their genetic material and possess the ability, via the enzyme reverse transcriptase, to make DNA from RNA and then insert their DNA into a host cell, are known as retroviruses. As we shall see later, they are of great value to the genetic engineer.

Other organisms

Although many people, if asked to divide life into two groups, might choose animals and plants, biologists prefer prokaryotes and eukaryotes. Prokaryotes include the various types of bacterium. Eukaryotes (from the Greek for 'true cell') comprise all the rest of life, namely fungi, plants, animals and all the one-celled organisms that seem to fall half-way between animals and plants. (Viruses are left outside this classification altogether.) The terms eukaryote is a useful one because all eukaryotes share a number of similarities, with respect to their chromosomes, that are not found in bacteria.

For a start, a eukaryotic cell contains a number of chromosomes held in a nucleus, whereas a bacterial cell lacks a nucleus and has only one chromosome. Further, eukaryotic chromosomes are more complex in structure than those found in bacteria. From the point of view of a genetic engineer, this is a pity, as it complicates the practice of genetic engineering.

One difference between eukaryotic chromosomes and those found in bacteria is simply the amount of DNA they contain. Indeed, the nucleus of a single human cell contains some 2 metres in length of DNA. Yet a typical human cell is only a hundredth of a millimetre in diameter! The remarkable packaging of DNA that is needed is illustrated in Figure 2.8. You can see that not only does a eukaryotic chromosome contain DNA that is highly folded, but it also contains a large amount of protein. These proteins help both in packaging the DNA and in protecting it.

A second important difference between the genetic material of bacteria and eukaryotes was discovered in 1977 when the exact sequence of bases in the gene that codes for the β-chain of haemoglobin was determined. Greatly to everyone's surprise, it turned out that the β-haemoglobin gene contains two regions of DNA whose base sequence does not correspond to the known amino acid sequence of β-haemoglobin. Further study revealed that the entire gene is transcribed into messenger RNA, but that some of the messenger RNA is then cut out and discarded before the messenger RNA leaves the nucleus. The chunks of RNA that are removed can be thought of as interruptions and are referred to as introns. Somewhat confusingly, the remaining bits of RNA – the

The DNA double helix....

....becomes coiled round
proteins to form a thread
like a string of beads

The proteins become
packed together to form a
more condensed thread

The thread becomes
folded....

....and folded again....

....into the condensed state
seen in the chromosome
during cell division

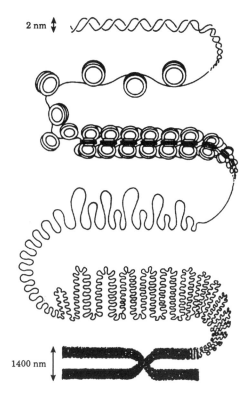

Figure 2.8. The structure of a eukaryotic chromosome, showing
how tightly the DNA is packed. There are a million nanometres
(nm) in a millimetre.

bits that actually end up coding for the polypeptide made by the
gene concerned – are called exons (Figure 2.9).

What is the purpose of these introns? One suggestion is that they
are parasitic bits of DNA, of no benefit to the rest of the cell but just
hanging in there ensuring they get reproduced generation after
generation. However, we now know that a mutation in one of the
introns of the β-haemoglobin gene can have serious consequences.
An abnormal form of this gene is known in which there is a gene
substitution in one of the two introns – a thymine instead of a
guanine. The presence of this thymine causes a mistake when the
time comes for the two introns to be cut out of the messenger RNA.
The net result is that although the first 29 amino acids of the

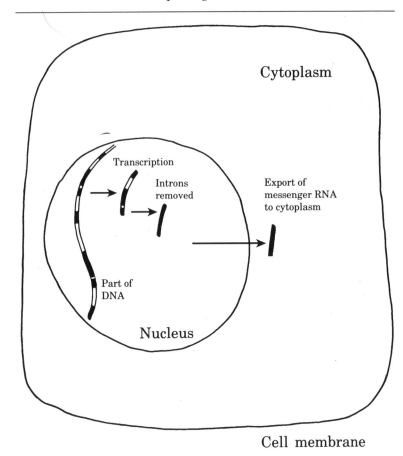

Figure 2.9. In a eukaryote, DNA contains introns. These introns are transcribed into RNA but are then cut out before messenger RNA leaves the nucleus and goes to the ribosomes in the cytoplasm where protein synthesis takes place.

polypeptide are correct, the next six are faulty and then the polypeptide stops. The polypeptide chain, therefore, has only 35 amino acids, whereas the normal β-haemoglobin chain has 141 amino acids. One can't be sure that introns always have such a role, but, in this case, a single faulty base in one of the introns causes a defective β-haemoglobin chain to be synthesised. A person who inherits this faulty gene from both their mother and their father suffers from thalassaemia major, a condition characterised by severe anaemia, growth retardation and a number of other problems.

In much the same way that introns occur, large amounts of what appear to be useless, or junk, DNA occur in eukaryotic chromosomes between functional genes. Again, it is unclear whether this DNA always has a function, or whether much of it is simply parasitic.

A number of other differences between bacteria and eukaryotes relate to the fact that in multicellular eukaryotes any particular cell makes only a small proportion of the polypeptide chains of which it is theoretically capable. The typical mammalian cell, for example, contains some 10 000 to 30 000 genes, yet it makes only a thousand or so different polypeptide chains. This is because different cells specialise and so need only transcribe a small proportion of their DNA into messenger RNA.

In addition to the differences between bacterial and eukaryotic DNA discussed so far, there are relatively subtle differences in chemical structure at the beginnings and ends of genes. These difference relate to the control, in eukaryotes, of messenger RNA and protein synthesis.

A final difference between bacteria and eukaryotes is that many of the polypeptide chains made by eukaryotes for export from the cell are modified before they leave the cell, for example they may have sugars added. Bacteria are unable to carry out these particular modifications, a fact that, again, complicates the life of the genetic engineer wishing to use bacteria to synthesise certain eukaryotic proteins.

The transfer of genetic material between organisms

It is important to emphasise that in nature sex is the way in which organisms transfer genetic material from one to another. When you think about it, sexual reproduction, whether in humans or any other organism, is functionally all about two individuals co-operating to produce a new individual with its own novel genetic make-up. As we discussed earlier (p. 23), this genetic make-up is derived equally from the mother and the father as a consequence of meiosis and the subsequent fusion of gametes in fertilisation.

In nature, sex only occurs between males and females *within* a species (or, extremely rarely, between very closely related species). Indeed, for most purposes a species can be defined as a collection of individuals able to breed only amongst themselves. Traditional biotechnology, as we discussed in Chapter 1, largely operates by selective breeding within species, though some hybridisation between closely related species also plays a part. However, the advent of genetic engineering has meant that these interspecific breeding barriers, evolved over countless millions of years, have, effectively, collapsed. Genes can be transferred almost at will from one organism to practically *any* other, from bacteria to plants, from plants to bacteria, from humans to mice and from humans to bacteria, to give just a few examples.

We are now in a position to discuss the techniques available for genetic engineering. It is convenient to look first at the basic principles of genetic engineering and then to look at the specifics of how genetic material can be moved between organisms.[2]

Basic principles of genetic engineering

Suppose that you, a genetic engineer, want one species to produce a polypeptide made by another species. For example, you might want a bacterium to produce human insulin so as to be able to collect and then give the insulin to people unable to make it for themselves, or you might want a crop plant normally susceptible to salt poisoning to produce a polypeptide made by a plant that naturally grows by the sea shore which helps the plant to be tolerant to salt spray. The basic procedure, using genetic engineering, involves the following two steps:

- Identify the gene that makes the polypeptide you are interested in
- Transfer this gene from the species in which it occurs naturally to the species in which you want the gene to be.

The first of these steps is actually more difficult than it may sound. After all, even a bacterium has hundreds of different genes, while animals and plants have tens of thousands.

Nowadays there are a number of ways of identifying the gene that makes the polypeptide in which you are interested. For example, suppose (quite a big supposition in itself!) you have a sample of the protein you are interested in (e.g. human insulin). Nowadays there are semi-automated techniques that enable you to work out the order of the amino acids in the protein. Having determined the amino acid order, you can then use the genetic dictionary in Table 2.1 to give you a pretty good idea of the sequence of bases in the messenger RNA responsible for the protein. (Note that you can't be 100% certain, as most amino acids are coded for by more than one triplet of RNA bases.) Having got a pretty good idea of the messenger RNA sequence, the DNA sequence of the gene concerned can then be deduced. (Note, though, that this only applies to the exons. There is no way of deducing the base sequence of the introns.) Finally, various techniques can enable you to track down this DNA. For example, you can synthesise some of the messenger RNA, the approximate composition of which you earlier deduced. Then you attach a radioactive marker to the messenger RNA. This messenger RNA will then bind to the gene concerned and the radioactive marker will tell you which chromosome is involved, and the approximate position of the gene on that chromosome.

It is with the second of these two steps that we are more concerned, namely transferring this gene from the species in which it occurs naturally to the species in which you want it to be. Two contrasting approaches can be used. The one involves the use of a vector organism to carry the gene; the other, called vectorless transmission, is more direct and requires no intermediary organism. Vectorless transmission is the simpler, and so we will consider it first.

Vectorless transmission

Remarkable as it may sound, one way of getting DNA into a new organism is to fire it in via a gun! This is known as biolistic (particle gun) delivery. The DNA is mixed with tiny metal particles, usually made of tungsten. These are then simply fired into the organism, or a tissue culture of cells of the organism. The chief advantage of this method is its simplicity, and it is widely used in research on the genetic engineering of plants. One problem, not surprisingly, is the damage that may be caused as a result of the firing process. A more intractable problem is that only a small proportion of the cells tend to take up the foreign DNA.

A second way of getting DNA into a new organism is by injecting it directly into the nucleus. This approach is quite widely used in the genetic engineering of animals. This method ensures a high proportion of the cells take up the foreign DNA.

A third type of vectorless transmission is known as electroporation. Here the cells to be genetically engineered are placed in a solution of the foreign DNA. A strong electric field is then applied. The electric field affects the membrane that surrounds each cell and leads to the DNA being taken up by the cells. Electroporation is used quite widely for the genetic engineering of plant and, to a lesser extent, fungal cells.

Recently, an even simpler way has been found of getting foreign DNA into plant cells. All you have to do is to shake together a test-tube containing water, the plant cells, the foreign DNA and tiny crystals of a common chemical called silicon carbide. The crystals punch very small holes in the plant cells, through which the foreign DNA can enter. There is then a good chance that the new DNA will be inserted into the plant's own DNA.

Vectors

A vector is an organism that carries genetic material from one species (the donor species) to another (the genetically engineered species). Genetic engineering by means of a vector involves three steps:

- Obtaining the desired piece of genetic material from the donor species
- Inserting this piece of genetic material into the vector species

- Infecting the species to be genetically engineered with the vector species so that the desired piece of genetic material passes from the vector to the genetically engineered species.

Actually, the third step may not always take place. For example, consider the synthesis of a human polypeptide chain by a bacterium. Here, only the first two of these three steps are needed. Step one involves obtaining the required piece of DNA from a human. Step two entails breaking open the DNA ring of a bacterial plasmid and inserting the required piece of human DNA into the bacterial plasmid. This procedure is called gene splicing.

Gene splicing relies on an important group of naturally occurring enzymes called restriction endonucleases. A given restriction endonuclease cuts the bacterial plasmid open at a specific site that is determined by the sequence of bases at that site. One of the most widely used restriction enzymes is known as *Eco*RI, so called because it is produced by a gene on an R plasmid in the bacterium *Escherichia coli*. *Escherichia coli*, or *E. coli* as it is generally known, is a bacterium found in the human gut and is very widely used in genetic engineering. *Eco*RI recognises the DNA sequence GAATTC and cuts the DNA between the G and A bases (Figure 2.10).

Most restriction enzymes cut DNA in such a way that one strand ends up with a few more bases than the other, as is the case with *Eco*RI. The result is that the double-stranded DNA has a short single-stranded sticky end. Sticky ends, as they are known, can join to complementary single-stranded sticky ends on other pieces of DNA. This property turns out to be extremely useful in genetic engineering. It means, for example, that if the same restriction endonuclease used to open up a bacterial plasmid is then used to obtain a piece of DNA from another species, the same sticky ends are produced. It is then a relatively straightforward matter to insert the DNA removed from the donor species into the bacterial plasmid. This process is illustrated in Figure 2.11. The enzyme DNA ligase is used to join up the sticky ends. Once in position, the foreign DNA replicates along with the rest of the plasmid every time the bacterial cell divides. The foreign DNA is said to have

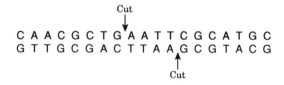

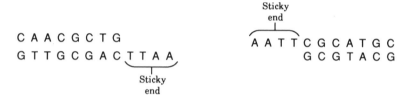

Figure 2.10. The action of the restriction endonucleae *Eco*RI. This finds the DNA sequence GAATTC and then cuts the DNA between the G and A bases.

been cloned and can be used for large-scale synthesis of the particular polypeptide for which it codes.

Of course, the enzymes needed for DNA splicing – restriction endonucleases and DNA ligase – did not evolve for the benefit of genetic engineers. They occur naturally in both bacteria and eukaryotes and have important functions in their own right. For example, they help protect cells against attack by viruses by trying to cut up the viral DNA before it has time to take over the cell. What genetic engineers have done is to exploit these enzymes, extracting them and using them as tools in genetic engineering.

Genetic engineering by means of a vector goes one step further than this. It involves infecting the species to be genetically engineered with the vector so that the desired piece of genetic material passes from the vector to the genetically engineered species.

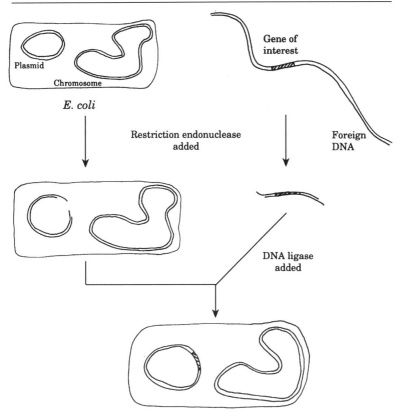

Figure 2.11. How to move a gene from one species to a bacterial plasmid using a restriction endonuclease to generate sticky ends and a DNA ligase to stick the sticky ends together.

An example is the infection of certain plants by genetically engineered forms of the bacterium *Agrobacterium*.

Agrobacterium is a soil bacterium that naturally attacks certain plants, causing a disease called crown gall disease in the process. *Agrobacterium* infects wounds and causes the development of swellings known as tumours. In 1977, it was found that the tumours were caused by the bacterium inserting part of one of its plasmids into the host DNA. This plasmid, known as the Ti (tumour induction) plasmid, was subsequently found to exist in a mutant form in which it still inserted part of itself into the plant DNA but did not result in crown gall disease.

The ability of *Agrobacterium* to insert part of its Ti plasmid into the DNA of its plant host has been harnessed by plant genetic engineers as follows:

- Insert the desired gene into the *Agrobacterium* plasmid, using a restriction endonuclease and ligase as already shown in Figure 2.11; this results in genetically engineered *Agrobacterium*
- Infect the plant you want to genetically engineer with the genetically engineered *Agrobacterium*. The Ti plasmid, even in its mutant non-disease-causing form, inserts the desired gene into the plant you want to genetically engineer.

Agrobacterium is a bacterium. Viruses can also be used as vectors in genetic engineering. We will close this chapter by looking at two ways in which viruses can be used in genetic engineering.[3]

Earlier (p. 29), we saw how retroviruses synthesise complementary DNA from their RNA and then insert this DNA into their host's DNA. This isn't so very different from the way in which the Ti plasmid of *Agrobacterium* operates, except that the Ti plasmid consists of DNA, so a retrovirus has to go through an extra stage: namely using the enzyme reverse transcriptase to synthesise DNA from its RNA. Retroviruses have been used in genetic engineering research on humans. For example, a number of diseases are caused by mutations in genes expressed in bone marrow cells. Bone marrow cells give rise to our blood cells. Retroviruses have been used in attempts to insert a functional copy of the faulty gene into these bone marrow cells. The aim is to ensure that all the blood cells that descend from these bone marrow cells are healthy.

Retroviruses are good candidates for this therapeutic approach as they have millions of years of experience at inserting their genetic material into that of a host. The general approach is to get a functional copy of the messenger RNA of the required human gene into the RNA of the retrovirus. On infecting the bone marrow cells, the retrovirus makes complementary DNA and inserts this, including the functional human gene, into the DNA of the bone marrow cells.

One problem with this approach, which limits the number of diseases for which it is being tried, is that retroviruses only infect dividing cells. Many human diseases, for example those of the nervous system, are not caused by mutations in dividing cells. A second problem is that, as so often in genetic engineering, there is no control presently available as to where the gene is inserted in the human chromosomes. Instead the retrovirus inserts its DNA, along with the desired gene, more or less randomly. This has two consequences. First, the new gene may not be as effective as it is when it is located in its normal place. This is because genes often work best only if they are situated close to certain other genes which help turn them on and off. The second, and more dangerous, possible consequence of the problem of random insertion is that the new gene may, by mistake, be inserted into what are known as tumour-suppressor genes. As their name suggests, these genes help prevent cancers. Disruption of the activity of a tumour-suppressor gene by the insertion of a gene through the activity of a retrovirus has been shown in monkeys sometimes to lead to the development of cancer.

For these reasons, researchers are experimenting with other viruses. For example, viruses known as adenoviruses are being used in attempts to insert functional copies of the gene that, in its faulty form, causes cystic fibrosis in humans. Adenoviruses, unlike retroviruses, do not integrate their genes into their host's DNA. This has both advantages and disadvantages. On the one hand, an obvious disadvantage follows from the fact that any descendants of the genetically engineered cells do not carry the functional cystic fibrosis gene. This means that once the genetically engineered cells die, the functional cystic fibrosis gene is lost with them. As a result, this approach is only likely to be effective if people with cystic fibrosis are treated with genetically engineered adenoviruses every few months. On the other hand, at least there is no risk of the virus inserting its genetic material into the host cells in such a way as to disrupt normal functioning or even to cause cancer.

Conclusions

We have seen in this chapter how the relatively recent discovery of the structure of DNA has led directly to the advent of genetic engineering, in which scientists can move bits of genetic material from one species to another. Much of what we have presented here is simplified, and knowledge and techniques in this field are advancing with great rapidity. Nevertheless, this chapter contains sufficient for us to be able, in Chapters 5 to 8, to look in more detail at genetic engineering in microorganisms, plants, non-human animals and humans. But, first, we will look in Chapter 3 at some general moral and ethical concerns that arise from these techniques and then, in Chapter 4, at the theology of genetic engineering.

3
Moral and ethical concerns

Biotechnology is the expansion, institutionalisation and misapplication of a particular scientific creed with the potential for the devaluation and exploitative manipulation of life

Talbot (1991)[1]

Biotechnology is ideologically neutral. Properly supported it can bring immense benefits to mankind, for it is infinitely adaptable to counter all sorts of unforeseen threats. If we cast it down through hostility or faintheartedness we shall all be losers.

Cross (1991)[2]

Why bother about moral and ethical concerns?

The former Archbishop of York, John Habgood, has referred to a 'growth industry in ethical discussion around biotechnology' and to a bibliography on the ethical and social implications of the human genome project that he had encountered which contained over fifteen hundred items.[3] This observation suggests two questions that should be addressed immediately in this chapter:

1. Do we need to swell the 'growth industry' and the bibliography yet further with the addition of this book?
2. More generally, does this amount of effort need to be directed towards discussion of such issues?

The answer to the first question is partly suggested by the specific focus of the bibliography mentioned by the Archbishop –

the human genome project. Much of the moral and ethical debate that has taken place has been one-dimensional in that it has been concerned with single, particular applications of genetic engineering, and actual or potential *human* applications have tended to attract most attention and interest. Attempts to examine more general concerns that arise horizontally across a whole spectrum of applications (to microorganisms, plants and animals as well as human beings) have been relatively rare; yet crucial questions need to be asked about whether or not *general* moral objections to the genetic modification of a plant can be transferred to that of an animal, or whether all applications raise distinctively different problems, which should be tackled separately. This overall mapping operation is a major objective of this book.

The second question, however, cannot be avoided. Why bother with this mapping operation at all? Two possible answers were briefly offered in Chapter 1, where it was claimed that no new scientific or technological development can claim immunity from ethical scrutiny and that moral and ethical concerns play an important role in influencing public attitudes. The first point needs no further elaboration but a little more must be said here about the second.

Concern is being increasingly expressed that the potential benefits of biotechnology may be lost if the new processes and products fail to gain 'consumer acceptance', and several studies of public attitudes have suggested that moral concerns act as a powerful influential factor here. One UK survey of attitudes towards technological innovations in food production,[4] for example, found that 70% of those questioned thought genetic engineering to be 'morally wrong', while a similar American study concluded:[5] '. . . the high potential for moral objection to biotechnology suggests that the issues may increasingly be framed in terms of basic values and fundamental beliefs. The belief that biotechnology is morally wrong had the strongest influence on acceptance of and attitudes about biotechnology'.

If moral concerns are so influential in this area, there must be a strong practical argument in favour not of trying paternalistically to persuade the public that genetic engineering is really a good thing but of encouraging moral judgments about it to be made on a rational and considered basis. More will be said about the educational dimensions of this task in Part 3.

What are moral and ethical concerns?

So far the terms 'moral' and 'ethical' have been used without explanation or definition, implying perhaps that their meanings are self-evident and synonymous. This is certainly a common assumption, but such assumptions do not always lead to clear thinking and rational argument. If genetic engineering is indeed a source of moral and ethical concern, as is frequently claimed, then we need to clarify what exactly constitutes 'moral and ethical concern' before the debate can proceed any further.

The terms 'moral' and 'ethical' are often used interchangeably in everyday language and, indeed, by some philosophers.[6] Nevertheless, distinctions can be drawn between them, and one of these will help in providing a structure for this investigation. It is important to emphasise, however, that the distinction that will be suggested here is by no means the only way in which the two terms can be distinguished, and that no attempt is being made to stipulate the 'correct' use of these terms or to claim that this is what they 'really' mean. There is certainly no consensus that 'moral' and 'ethical' should be used in the following way, and what matters in linguistic questions of this kind is not the particular labels themselves but the realisation that two different things with distinctive features can be distinguished and that greater clarity will be achieved if they are not confused. Given these provisoes, a possible distinction between 'moral' and 'ethical' will now be outlined and thereafter adopted for the purposes of this book.

> **Moral**. Everybody (except perhaps the psychopath) can be said to have moral views, beliefs and concerns, to the effect that certain things are right or wrong and that certain actions ought or ought not to be performed. These views may refer to virtually any subject; a person may feel that it is wrong to hunt foxes, to make jokes about the Royal Family, to work on Sundays, or to smack

children. Such 'moral concerns' may result from a lot of
deliberation and reflection, or from very little; they may
be firmly grounded in a consistent set of carefully
considered rational principles, or they may not; their
justification may have been consciously analysed, or it
may not. Many of our moral views are probably held
almost unthinkingly, perhaps as a result of our
upbringing. We may just 'feel' that certain things are right
or wrong; we have a 'gut reaction' about them; and that
may be the sum total of some people's 'morality'.

Ethical. Ethics is normally thought of as a narrower
concept than morality,[7] and it can be used in several
different, though related, senses. The most general of
these: '. . . suggests a set of standards by which a
particular group or community decides to regulate its
behaviour – to distinguish what is legitimate or acceptable
in pursuit of their aims from what is not. Hence we talk of
"business ethics" or "medical ethics"'.[8]

More technically, ethics can also refer to a particular
branch of philosophy – 'moral philosophy'. There is
disagreement among philosophers about the precise
scope, function and methodology of ethics, but this is not
the place for a detailed discussion of such issues. For our
present purposes, we should note that one central task of
ethics is usually taken to be a critical investigation of the
fundamental principles and concepts that are used in
moral debate. Ethics tries to analyse and clarify the
arguments that are used when moral questions are
discussed and to probe the justifications that are offered
for moral claims. So ethics in this sense is a critical
'second-order' activity which puts our 'first-order' moral
beliefs under the spotlight for scrutiny, but this is not an
activity reserved solely for professional philosophers: 'In
so far as the man in the street thinks critically about his
own moral views or those of others, or ponders on their
justification, or compares them with rival attitudes, to that
extent he is a moral philosopher'.[9]

This way of distinguishing between 'moral' and 'ethical', therefore, shows how essential it is to 'unpack' the apparently straightforward statement that genetic engineering is a source of moral and ethical concern; for to call something a *moral* concern does not necessarily mean that it is of much *ethical* significance. Many people in the past felt moral concern about old ladies who lived alone with black cats, but what ethical validity did such feelings have? Genetic engineering may be rather a similar source of moral concern to some people today. The survey of public attitudes in the UK towards new technological innovations in food production,[4] already referred to, found not only that 70% of those questioned thought 'genetic engineering' to be 'morally wrong', but also that 62% thought it 'unnatural' and 27% thought it 'frightening.' If asked, therefore, people will express general *moral* concern and unease about genetic engineering (as they would have done about old ladies with black cats), but this does not tell us whether they have done any *ethical* thinking about the issues.

How can moral and ethical concerns be evaluated?

The approach to be adopted in this book is based upon the above distinction between 'moral' and 'ethical'. Those aspects of genetic engineering that appear to give rise to most moral concern will be described, and the various ways in which that moral concern is expressed will be examined and subjected to ethical scrutiny, which will analyse the concepts used and the principles implied. This approach is intended to identify those issues of most ethical significance and to distinguish them from those of less substance. This analysis may also help to defuse some of the controversy surrounding genetic engineering, which often generates more heat than light. As one writer puts it: 'A significant part of the current debate can be traced to differences over moral principles. Also, unfortunately, there has been much unnecessary debate generated by careless moral reasoning and a failure to attend to the logical structure of some of the moral arguments that have been advanced'.[10]

Differences over moral principles, however, cannot be resolved by a simple appeal to 'the facts', for sets of facts whether about genetic engineering or anything else can never *prove* something to be morally right or wrong. Yet this does not mean that facts are unimportant or irrelevant when considering ethical questions: one cannot make informed ethical judgments about *x* without knowing the relevant facts about *x*, which is why this book deals with the science as well as the ethics of genetic engineering.

Ethics cannot, then, offer conclusive *answers* about the rightness or wrongness of genetic engineering for ethics cannot provide final 'proof' of this kind. One cannot prove that the hungry ought to be fed in the same way that one can prove that lack of food causes death. Moral judgments may be argued for or against, criticised or defended, and shown to be more or less rational and informed, but their rightness or wrongness can never be comprehensively established.

What ethical methods can be used to evaluate moral concerns? We know in principle how to set about evaluating a scientific claim that a particular genetic combination produces particular attributes in pigs, but how do we evaluate moral claims that it is wrong to apply genetic engineering to pigs? This raises vast and complex questions about the nature of ethics and the logic of moral judgments, and, as mentioned above, this has long been an arena for philosophical disagreement and competing theories. Some of these theories (e.g. utilitarianism) will need to be discussed where appropriate, but our aim is not to produce a comprehensive ethical textbook, of which there are more than enough already, but rather a framework for assessing the concerns that these technological innovations are generating. At this point, therefore, all that can usefully be said in general terms about the ethical methods that will be used is that they will emphasise two objectives:

1. The analysis and clarification of the key concepts that tend to be used when these concerns are expressed (e.g. that genetic engineering is 'unnatural')
2. The uncovering and probing of the underlying principles upon which the concerns appear to be based (e.g. that it is wrong for scientific research to take risks).

These two approaches will feature prominently in the following chapters. In Parts 2 and 3, a wide range of specific moral concerns will be examined, arising from particular applications and implications of genetic engineering. There are also, however, moral concerns of a more general and fundamental nature relating to the technology itself and to all or most of its applications, and these will be investigated in the remainder of this first section.

Extrinsic and intrinsic concerns

The moral concerns expressed about genetic engineering may, for convenience, be divided into two basic categories, to be labelled 'extrinsic' and 'intrinsic'. Genetic engineering may for a variety of reasons be thought to be either intrinsically wrong *in itself* or extrinsically wrong *because of its consequences*. This important distinction can be applied to a large number of moral issues and can often help in identifying the precise grounds of any moral concern. Confusion can quickly arise if the distinction is not drawn. A debate about the justifiability of abortion, for example, will not get very far if the participants fail to realise that one (intrinsic) set of arguments – that abortion is murder and, therefore, always wrong in itself – is of a different logical form from and so cannot be countered by another (extrinsic) set of arguments – that the consequences of allowing certain pregnancies to continue are sometimes morally unjustifiable.

Intrinsic arguments cut deeper than extrinsic ones. If abortion, genetic engineering or anything else is thought to be intrinsically wrong, no further considerations are morally relevant, for nothing can reverse that intrinsic wrongness; consequences do not have to be taken into account. Intrinsic arguments about the wrongness of something also focus attention upon its precise nature and its distinguishing characteristics. What distinguishes abortion from other medical procedures that may result in death? What distinguishes genetic engineering from traditional selective breeding? Wherein exactly does the alleged intrinsic wrongness lie?

Intrinsic arguments also have a more specific, clearcut target than extrinsic ones. If I feel that abortion is intrinsically wrong, then it is clearly the act of abortion *per se* that I see as a moral concern; but if I feel moral concern about driving fast cars or owning a pit bull terrier, it is not normally those activities in themselves that I am objecting to, but their possible extrinsic effects and consequences, which could also result from all sorts of *other* activities.

This feature of extrinsic concerns raises a major problem. To claim that any activity or process will have undesirable consequences is to make *predictions* about future events. But predictions may be accurate or inaccurate, and no conclusive proof can ever be provided that a particular set of events will inevitably occur in the future – a point which will be re-emphasised shortly. Extrinsic concerns must, therefore, always be in this sense provisional: they carry weight only in proportion to the likelihood of the predicted consequences actually occurring. So to appraise the validity of these concerns becomes in part a technical matter of trying to establish what really is most likely to happen, requiring in this case the specialist expertise and judgment of the biologist.

Fortunately, however, it is not essential to be a professional biologist in order to reach a reasonably informed ethical judgment about the consequences of genetic engineering. Professional scientists are not the only people capable of assessing the moral implications of scientific developments. We do not need to be highly qualified nuclear physicists in order to form considered ethical judgments about the standards of care and responsibility required in the handling of nuclear materials; the common-sense understanding of the intelligent layperson is quite adequate to appreciate the moral irresponsibility of running avoidable risks with dangerous products and processes.

Ethical questions, then, can still be directed at these extrinsic concerns despite their technical nature. Indeed it is essential that they should, for the following reasons:

1. Even if agreement is reached about likely consequences (which, as we shall see, is rare) this does not automatically answer the moral and ethical questions. We still have to ask what is good or bad, right or wrong,

about those consequences and examine the moral claims and assumptions surrounding them.

2. There is never just *one* consequence to any activity but a whole set of consequences, often occurring at different times. Building a new motorway, for example, could produce an infinite number of consequences, which might include improving the traffic flow from point A to point B, increasing air pollution at point C and reducing it at point D, destroying the habitat of species E and providing a new one for species F. The consequences of any activity, therefore, cannot simply be morally approved or condemned *en bloc*, for they will often produce conflicting advantages and disadvantages.

3. Consequences, then, have to be *weighed* and *compared* against each other, and this cannot be a matter of purely factual assessment. Attempts to estimate the likely costs and benefits of an activity can, of course, be made on a straightforward financial basis, but this does nothing to address the moral issues. (A financial assessment of this kind could have been made in deciding the method of extinction to be used in Nazi concentration camps.) Ethical judgments have still to be made about the *value* or *priority* to be placed upon different possible costs and benefits produced by different possible consequences. This can be a highly complex matter; how, in the above example, are the various costs and benefits of building a motorway to be valued and prioritised?

Having distinguished between extrinsic and intrinsic concerns and noted some general problems involved in evaluating the former, we can now proceed to examine the most obvious example of an extrinsic concern – that genetic engineering is risky.

Extrinsic concerns about risk and safety

The most commonly expressed general concern about the possible consequences of genetic engineering is that these might be risky or even catastrophic. Eloquent critics of genetic engineering, such as Jeremy Rifkin in the USA, have concentrated much of their fire upon this concern, as the following two extracts from Rifkin's work illustrate:

> Now please just imagine introducing thousands of genetically-engineered organisms: bacteria, viruses, plant strains and animal breeds in massive volumes for commercial purposes. Sheer statistical probability, my friends, suggests that they are not all going to be safe.[11]

> Whenever a genetically engineered organism is released there is always a small chance that it too will run amok because, like exotic organisms, it is not a naturally occurring life form. It has been artificially introduced into a complex environment that has developed a web of highly synchronised relationships over millions of years. Each new synthetic introduction is tantamount to playing ecological roulette. That is, while there is only a small chance of it triggering an environmental explosion, if it does, the consequences can be thunderous and irreversible.[12]

Concern about particular types of risk will clearly vary depending upon the particular application of genetic engineering that is under discussion, and Rifkin is here referring mainly to agricultural applications involving plants and animals. We shall consider the possibility of specific risks in more detail in Part 2, where different applications of the technology will be examined, but some more general points about risk and safety and their ethical significance need to be made at this point.

One possible manoeuvre is to by-pass the issue completely by arguing that risk and safety are not in themselves moral or ethical matters. There is indeed something to be said for this viewpoint, which challenges the widely held (and often unexamined) assumption that any discussion of the morality and ethics of genetic engineering must immediately focus upon questions of risk and safety. A little reflection, however, shows that the connection is less obvious. The safety of a product, process or activity is, at least on the face of it, an empirical matter to be determined by experiment and experience. Whether or not a toadstool is safe to eat is not an ethical question. It is more risky to drive on motorways on wet Friday evenings than on fine Sunday mornings, but this is a statistical fact rather than a moral issue. Some activities are inevitably more risky than others, though none can be totally risk-free, and it does not follow that low-risk activities (e.g. snoozing in front of the television) are morally superior to high-risk ones (e.g. rescuing children trapped in burning buildings).

It would, however, be difficult and somewhat short-sighted to maintain that questions about risk and safety can have no moral dimension. Risk and safety become matters of moral concern when they raise further questions about *responsibility*, *accountability* and *justifiability*. Moral concern is appropriate when irresponsible and unjustifiable risks are thought to be taken, which may result in harm to innocent parties: joyriding at high speed in a stolen car arouses moral concern to a far greater extent than driving a racing car in a Grand Prix. Furthermore, the *assessment* of risks may also be shown to have an ethical component, a point emphasised in a report on genetic engineering by a Swedish Government Committee, which is one of the few such documents to discuss explicitly the relationship between risk and ethics:

> The Committee takes the view that the assessment of risks should be part of the ethical analysis, since it is ethically false to base a decision on poor foundations if the decision can be postponed until the foundations have improved. It is also ethically unacceptable to assert that the foundations for a decision are better than they are. Thus the open presentation of facts and viewpoints is important.[13]

The question of risk cannot, therefore, be ignored in any ethical investigation of genetic engineering, and we need to look a little more closely now at some of the general concerns about risk that have been generated by the development of this technology.

The early history of the development of genetic engineering techniques was dominated by disputes about safety, particularly in the USA. These disputes have been graphically documented in great detail in a book by Krimsky, which describes the 'social history of the recombinant DNA controversy'.[14] Basically the fear in the 1970s was, and to some extent still is, that genetically engineered organisms could escape or be deliberately released from the laboratory into the environment with unpredictable and possibly catastrophic consequences. A particular concern was that the bacterium *Escherichia coli* was being used in experimental work and that, as this bacterium resides naturally in the human gut, genetically engineered variants of it might cause an uncontrollable spread of disease outside the laboratory. Other ecological disasters were hypothesised in the event of modified microbes escaping and 'upsetting the balance of nature'. An American report justifying new regulatory legislation in 1977 summed up the possible dangers as follows:

> Foreign DNA in a micro-organism may alter it in
> unpredictable and undesirable ways. Should the altered
> micro-organism escape from containment, it might infect
> human beings, animals or plants causing disease or modifying
> the environment. Or the altered bacteria might have a
> competitive advantage, enhancing their survival in some niche
> within the ecosystem.[15]

During the 1970s, increasingly stringent regulations were introduced, and in Japan and Holland genetic manipulation research was totally banned. During the 1980s, however, these regulations were gradually relaxed as confidence increased in the view that modified microbes are unlikely to pose significant ecological threats.

The complex issue of regulation will be considered in more detail in Part 2, as it is clearly of central importance in trying to determine whether irresponsible and unjustifiable risks are being taken in genetic engineering. But if the possibility of catastrophic consequences may exist, as envisaged by Rifkin and others, is it perhaps irresponsible and unjustifiable to proceed with this technology at all?

It has been suggested that these consequences might include uncontrollable diseases, as already mentioned, global drought (caused by the 'ice-minus' organism designed to protect crops against frost) or the spread of indestructible weeds made resistant to pests and herbicides. Further fears have been voiced that a loss of genetic diversity among plants and animals could result, making them less resilient and so more vulnerable to various forms of attack in the future. More generally, some have claimed that genetic engineering could represent the first step on a slippery slope that will lead inexorably to a nightmare programme of universal eugenics: 'By continuing along this road, we could end up reducing the human species to a technologically designed product'.[16]

The risks envisaged here are clearly of such a catastrophic nature that no one (with the exception of the stereotypical 'mad scientist' of sensational fiction) would feel justified in turning a blind eye to them. So can we cut short our ethical investigation at this point by accepting that such risk-taking is irresponsible and unjustifiable on the basis of a principle to the effect that any activity which could lead to catastrophic consequences ought not to be undertaken? Unfortunately, this simple and apparently responsible conclusion becomes less convincing when we look more closely at the principle on which it depends. The main objection is well summarised in a detailed analysis by Stich:

> Once the principle has been stated, it is hard to believe that anyone would take it at all seriously. For the principle entails that, along with recombinant DNA research, almost all scientific research and many other common-place activities having little to do with science should be prohibited. It is, after all, at least logically possible that the next new compound synthesized in an ongoing chemical research program will turn out to be an uncontainable carcinogen many orders of magnitude more dangerous than aerosol plutonium. And, to vary the example, there is a non-zero probability that experiments in artificial pollination will produce a weed that will, a decade from now, ruin the world's food grain harvest.[17]

Stich's point about 'non-zero probability' is one that is often blurred in the presentation of 'safety issues' by the media: it is not uncommon to hear or read interviews in which scientists and politicians are asked, 'Is X really safe?', and are then accused of equivocation if they refuse (as they should) to give a straightforward yes or no answer. The simple logical truth, which is of some ethical relevance here, is that it is impossible to *prove* that a particular event will or will not happen in the future. No activity or process can ever be guaranteed to present no risk whatever and to be 100% 'safe'; genetic engineering is no exception to this logical rule, as Godown emphasises in a paper on the science of bio-technology:

> Can science tell us for instance what will be the result of creating and releasing a novel organism from which a single gene has been deleted? Could it ever be flatly stated on the basis of scientifically established facts that there is no possibility of anything going wrong when a genetically engineered organism is deliberately released? The answer is obviously no . . . One cannot prove a universal negative and it is silly to try.[18]

The claim that genetic engineering is potentially hazardous, therefore, carries less weight than it at first appears to do, once it is realised that *everything* is potentially hazardous. Further confusion can be caused by the fact that the term itself, 'potentially hazardous', is highly ambiguous, as Krimsky's analysis shows. 'X is potentially hazardous to P' may be interpreted as:

1. X can harm P under conditions $C_1 \ldots C_n$.
2. It is not known that X cannot harm P under some set of conditions or another.
3. There is some evidence that X may be harmful to P.
4. There is a finite probability that X can harm P.
5. There is a posited scenario of events such that X harms P, where the scenario has neither been confirmed nor disproved.[19]

But is it perhaps mere sophistry to place too much weight on such logical and linguistic points? Critics like Rifkin argue that the risks involved here are of such a *level* as to make the further development and application of genetic engineering irresponsible; it is the particular and peculiar risks associated with these techniques that make them morally unjustifiable. Clearly the issue here is partly one of technical assessment, which will need to be considered in greater detail in Part 2 in the context of specific applications and examples.

More general and fundamental questions, however, can be raised at this point about the responsibilities and obligations of scientists, and the relative value to be placed on scientific progress and the pursuit of new knowledge. For example, it can be argued on the one hand that 'the fundamental ethical posture of science should be to do no harm, which implies reducing the risks to a negligible factor regardless of the anticipated benefits';[20] and, on the other hand, that 'it is morally wrong as well as politically dangerous to place restrictions on intellectual activities'.[21]

The main problem here is that it is difficult if not impossible to determine what the 'safe' option really is. What constitutes 'risk-avoidance' in this case? Excessive caution does not necessarily remove the risk of future catastrophes. By banning research and development in any new technology that is thought to involve risks, we may run the risk of failing to produce an innovation that will be desperately needed in some future, unforeseen crisis. The history of science has proved to be highly unpredictable, and there can be no guarantee that 'playing safe' by abandoning research and development in genetic engineering will not deny us a technique or product that may *prevent* an environmental disaster in 50 years time.

In deciding whether unjustifiable risks are being taken in genetic engineering, therefore, a balance has to be struck between the paralysis of extreme caution and the irresponsibility of uncontrolled experimentation. Since safety can never be totally 'proved', judgments have to be made about the likelihood of possible consequences and the relative value, desirability and priority to be assigned to these. This is an area where it is essential that science and ethics proceed hand in hand.

But how exactly can ethical judgments be made about such issues as risk and safety? Dangerous outcomes are but one of the possible consequences of genetic engineering; others might include the alleviation of world hunger or the exploitation of economically vulnerable individuals and countries. Any ethical assessment of the technology in terms of its possible consequences, then, will have to attempt to weigh the potential costs and benefits in some way. These costs and benefits cannot, however, be purely or even predominantly economic or financial, for ethics cannot be equated with accountancy. The costs and benefits must relate to a wider range of considerations, and while there is no philosophical consensus over the precise nature and extent of these, they are usually thought to include such things as the welfare, interests, rights and needs of human beings (and perhaps also of other animals).

Ethical theories that focus upon consequences in this way are usually labelled (unsurprisingly) 'consequentialist', and the best known of these is utilitarianism, which in its original and simplest form maintained that actions are right or wrong in proportion to the total amount of pleasure or pain that they produce, thus making the greatest happiness of the greatest number the ultimate ethical criterion. There are, however, some obvious objections to this approach, including:

- How in practice is this ethical arithmetic to be carried out?
- How is unfair treatment of innocent individuals to be avoided?

These problems are well illustrated in the case of genetic engineering. In the first place, even if the likely consequences can be accurately predicted (which is a tall order in itself), how exactly are the predicted costs and benefits to be weighed in terms of overall welfare and happiness? How, for example, is the possibility of increased food supplies to be weighed against the possibility of an invasion by indestructible weeds? A *further* ethical principle seems to be needed to resolve such issues and how is *that* in turn to be justified? And in the second place, how are the interests of particular individuals (e.g. small farmers) or perhaps of animals to be protected if the ethical arithmetic indicates a course of action that is likely to benefit the majority but cause serious harm to members of such minority groups?

Clearly there are no easy answers to such questions. Most philosophers would now agree that the original formulation of utilitarianism is too crude to cope with the complexity of moral concerns, and a number of attempts have been made to produce more refined versions of the theory. We do not need, however, to become further embroiled in the intricacies of utilitarianism at this point, as enough has been said to indicate the general problems involved in an ethical evaluation of genetic engineering in terms of its possibly risky consequences. These problems of ethical arithmetic will be illustrated in more detailed examples in Part 2.

But for some people the possible consequences of genetic engineering will be seen as irrelevant. The technology itself can be thought to be intrinsically wrong, regardless of its results, as was pointed out earlier in this chapter. Having reviewed some of the issues raised by a common extrinsic concern about risk and safety, therefore, we can now turn to the more sweeping claim that genetic engineering is wrong in itself. Two examples of this claim will be examined, namely that genetic engineering is unnatural and that it shows a lack of respect for nature.

Intrinsic concerns about unnaturalness

Many people feel that genetic engineering is in some way unnatural, as the survey already mentioned indicates. For some, this may not be an intrinsic concern but rather another way of expressing fears about possibly risky consequences. The argument here would be that 'Nature knows best' and that the natural world as we know it is the result of a long evolutionary process with a well-established track record, whereas genetic engineers are gambling with their unproven introductions. We do not need to say more about this argument at this point (though it will appear again in Part 2) as the general issue of risk and safety has already been considered in this chapter. The more distinctive and fundamental concerns about the unnaturalness of genetic engineering are, however, of an intrinsic kind.

In many cases, there may well be a religious foundation for such feelings: Nature, representing the perfect work and will of God, is being interfered with by this new technology. Religious views about genetic engineering, however, can cover a very wide spectrum, and the complexity of the issues raised deserves separate treatment. The following chapter will accordingly be devoted to specifically religious and theological questions, including those concerned with the issue of 'natural law'.

Fears about unnaturalness need not rest upon a religious basis. It seems unlikely that the 62% of respondents in the survey quoted,[4] who thought genetic engineering unnatural, for example, were all basing their judgment upon explicit religious arguments.

So what is the basis of this concern and what ethical principles does it imply? Reduced to its simplest form, the argument seems to be as follows. 'Nature and all that is natural is valuable and good in itself; all forms of genetic engineering are unnatural in that they go against and interfere with Nature, particularly in the crossing of natural species boundaries; all forms of genetic engineering are, therefore, intrinsically wrong'.

The various elements in this argument need to be separated and examined in the light of two basic questions: what is meant by 'natural' and 'unnatural', and what is good about being 'natural' and bad about being 'unnatural'?

What is meant by 'natural' and 'unnatural'?

Before the above argument can even get off the ground, we have to be able to identify and agree about what is to *count* as 'natural' and 'unnatural'. This is no easy task in a world where we are offered natural beef, natural toothpaste, natural birth-control and a host of other allegedly 'natural' products and processes.

Depending on the context in which it is used, the word 'natural' may mean 'usual', 'normal', 'right', 'fitting', 'appropriate', 'uncultivated', 'innate', 'spontaneous', and no doubt many other things as well. Perhaps most commonly, 'natural' is contrasted with 'artificial' or 'man-made', but on the basis of that distinction practically every element of our modern Western life-style is 'unnatural'. Nor can more traditional products and processes avoid such a charge of 'unnaturalness', for the progress of civilisation has been largely dependent upon our 'interference with Nature'. Yet if every domestic or farm animal, every garden plant or agricultural crop, every item of food or clothing is thought of as unnatural because it interferes with Nature, as logically it must, then the concept of 'unnaturalness' surely becomes so broad as to be meaningless.

The more specific and serious charge of 'unnaturalness' that has been levelled against genetic engineering, however, is that it breaches natural species boundaries and violates the natural integrity of species:

> Genetic engineering makes it possible to breach the genetic boundaries that normally separate the genetic material of totally unrelated species. This means that the *telos*, or inherent nature, of animals can be so drastically modified (for example by inserting elephant growth hormone genes into cattle) as to radically change the entire direction of evolution, and primarily towards human ends at that. Is that aspect of the animal's *telos* we refer to as the genome and the gene pool of each species not to be respected and not worthy of moral consideration?[22]

A biologist might try to refute the argument that it is unnatural to breach the genetic boundaries between species on a number of grounds. For a start, such a view fails to realise that the theory of evolution, on which all our understanding of the nature of species is based, *requires* that species change over time. Every species alive today is believed to be the direct descendant of the early single-celled species that existed over three billion years ago. Species that currently exist have passed through many, possibly hundreds, of separate speciation (formation of a new biological species) events. In other words, species are not static; their genetic composition naturally changes over time.

Further, a view of evolution that assumes that species remain genetically isolated from one another is out of date. We now realise that a number of distinct processes allow the movement of genetic material from one species to another. Certain viruses, for example, carry genetic material between species. Equally, many bacteria have mechanisms that allow them to take up genetic material from other species and then integrate it into their own. In other words, for many species, their *telos* includes the ability to cross species barriers.

Finally, a biologist might point out that no single universally accepted criterion of the term 'species' exists. The word is used by different biologists in a number of distinct ways.[23] At one extreme, the term is employed to refer to a group of individuals that are able to breed only amongst themselves and not with members of other 'species'. Given this definition, it is simply tautologous to assert that it is unnatural to cross species barriers. At the other extreme, and this is the way in which, in practice, most biologists use it, the term species refers to a group of individuals that look very similar to one another or, to be more formal, that share certain morphological criteria that render them distinct from other 'species'. In this way, mallards can be distinguished, on such criteria as visual appearance and behaviour, from shellducks, teal and other species of waterfowl. A biologist with this understanding of what a species is will not be particularly bothered by occasional instances of breeding between different species of waterfowl.

These arguments serve to caution us against accepting unquestioningly the objection that genetic engineering is unnatural because it involves the crossing of species boundaries. Nevertheless, such reasoning can appear more convincing than it really is. For a start, the argument that species are not naturally static with respect to their genetic composition fails to acknowledge the remarkable slowness with which evolution normally acts. It usually takes more than a million years for one species to evolve into another; only a handful of examples are known in which animal species have evolved into new ones within less than 5000 years. Genetic engineering operates on a timescale of months or a few years, not millennia or aeons.

Further, it is important not to exaggerate the incidence in nature of the exchange of genetic material between species. True, such exchange is important in soil bacteria and can be mediated, in some other organisms, by certain viruses. Nevertheless, the movement of genetic material between species is, for most species, almost certainly both rare and of limited significance. This is, of course, not the case with genetic engineering. The genetic engineer is able to manipulate conditions so that these transgressions of species boundaries are both frequent and of major importance.

Finally, the fact that a genuine academic debate exists as to the precise meaning of the term 'species', though worth noting, is of little significance in the context of a debate about the naturalness or otherwise of the movement of genetic material between species.

Whatever biological definition of a species is accepted, most genetic engineering involves the movement of genetic material between species. It is fair to conclude that in the above respects most genetic engineering is 'unnatural'. However, the argument about 'unnaturalness' faces other philosophical objections.

What is good about being 'natural'?

The argument under consideration assumes that whatever is 'natural' is good and whatever is 'unnatural' is bad, but is this assumption warranted? A 'natural' event, product, process or tendency (however defined) is not automatically good or desirable. Many 'natural' substances are harmful; many of our 'natural' tendencies and reactions, such as jealousy and aggression, are not normally thought morally praiseworthy; many 'natural' events, such as earthquakes and hurricanes, create destruction and suffering and are, indeed, usually labelled 'natural' disasters; many 'natural' organisms cause pain, disease and death. As the theologian Don Cupitt points out, Nature can be seen as a 'kindly mother, lovely in every aspect' but also as 'wild, chaotic and pitiless'.[24] Darwin, the founder of modern evolutionary theory, shared the latter view, lamenting the 'clumsy, wasteful, blundering, low and horribly cruel works of nature'.

To assume that we can simply deduce what is morally right and wrong from certain facts about the world and about Nature is to commit what philosophers have called the 'naturalistic fallacy,' often translated as 'You can't get an ought from an is!' The logical point at issue here is really a very simple one: simply because something happens in nature does not mean that it is right or good, that it should be preserved or protected. A specific example of the naturalistic fallacy can be found in the argument about breaching natural species barriers. Even if these barriers unequivocally can be identified (which appears unlikely), their mere existence provides no clear ethical directives about what *ought* to be done about them. The River Thames is a 'natural' barrier between Surrey and Middlesex, but that geographical fact tells us nothing about whether it is morally right or wrong to cross from Surrey to Middlesex.

Claims about the 'unnaturalness' of genetic engineering, there-fore, do not appear to have much ethical significance, resting as they do upon unclear language and unsound reasoning. An argument to the effect that genetically engineering a drought-resistant strain of plant, for example, is 'unnatural' and, therefore, wrong would hardly stand up to much critical scrutiny. There are, however, more sophisticated versions of the 'unnatural' argument that some believe to carry greater ethical weight. These can take several forms but all focus in some way upon a lack of respect, which modern biotechnology is thought to embody.

Intrinsic concerns about disrespect

What kind of disrespect might genetic engineering be accused of exhibiting, and towards what? There are two main arguments to examine here, each concerned with aspects of our relationship with the natural world.

The reductionist argument

This view has again been eloquently propounded by Jeremy Rifkin:

> Already researchers in the field of molecular biology are arguing that there is nothing particularly sacred about the concept of a species. As they see it, the important unit of life is no longer the organism, but rather the gene. They increasingly view life from the vantage point of the chemical composition at the genetic level. From the reductionist perspective, life is merely the aggregate representation of the chemicals that give rise to it and therefore they see no ethical problem whatsoever in transferring one, five or even a hundred genes from one species into the heredity blueprint of another species. For they truly believe that they are only transferring chemicals coded in the genes and not anything unique to a specific animal. By this kind of reasoning, all of life becomes desacralized. All of life becomes reduced to a chemical level and becomes available for manipulation.[25]

This kind of claim still depends heavily upon assumptions about species boundaries that have already been shown to be highly debatable. When those assumptions are challenged, the reductionist argument as such loses much of its force. Of course, it is possible for any researcher to come to adopt a reductionist view of life as a result of his or her professional work, but that danger is just as great for the economist or social scientist (in terms of viewing human beings as mere statistics) as for the molecular biologist. In any case, the possible psychological effects of pursuing a particular subject cannot constitute a serious objection against the subject itself. Should historical research, for example, be banned on moral grounds because a few historians may become obsessed by the macabre details of public executions? If a genetic engineer finds that he is starting to see his wife, his children, his dog and his flower bed 'reduced to a chemical level', the answer is surely not to condemn or outlaw genetic engineering but for that unfortunate individual to seek psychiatric help.

There is, however, a broader dimension to the reductionist argument that deserves to be taken more seriously, namely a concern that genetic engineering shows a certain lack of respect for the environment. This leads us to the second argument to be examined.

The holistic argument

This set of views applies to a far wider range of issues than those raised by genetic engineering and embraces many ideas that are often loosely labelled 'holistic', 'ecological' or 'environmental'. Here is not the place to analyse these ideas in great depth, other than to note that they include claims and theories about the interdependence of all life-forms in a complex, self-regulating 'biotic community', and the consequent extension of moral rights and moral value to the non-human world. A new 'environmental ethic' is thus implied: 'People who really value wilderness and natural systems will not think it morally permissible for the last people on earth (who know they are the last people) to set about destroying the plant and animal species of their world. . . . People with the new values will disapprove of certain ways of using natural systems and living creatures'.[26]

Would these 'ways of using natural systems' include genetic engineering? Do these techniques in some way lack respect for the 'biotic community'? The World Council of Churches report *Integrity of Creation*, for example, asserts that they are associated with, 'a world view that does not respect humanity's dependence on the earth as mother and as the source of life and nourishment'.[27]

But what exactly is 'respect' and how do we display a lack of it? The importance of respect as an ethical principle was underlined by the German philosopher Kant, who argued that respect required treating others as *ends*, never only as *means*.[28] So to use another being *instrumentally*, entirely for one's own purposes, without taking any account of the other's interests, involves a lack of respect and is morally wrong.

Can genetic engineering be accused of lacking this kind of respect? The following points need to be taken into consideration.

First, Kant's account of respect makes it clear that we should never treat another being 'as means *only*'. In other words, as a modern philosopher has put it: 'In Kant's view at least, treating another life-form as an instrument is not incompatible with showing it respect. There is a distinction between using another creature's ends as your own – which is acceptable – and disregarding that other creature's ends entirely – which is not'.[29]

This point is perhaps best illustrated by examples of our possible relationships with other animals. Some forms of so-called factory farming appear to use animals purely as a means to increased food production, considering the animals' 'ends' only in so far as that will result in a better 'product' or higher 'productivity'. Less intensive forms of farming, however, while still using animals for human ends, may also be able to show some respect for those animals' ends (i.e. their 'natural' wants and tendencies, and even perhaps their individuality). This distinction is also reflected in the recent trend towards 'welfare labelling' of animal food products (e.g. the RSPCA's 'Freedom Food' campaign), the rationale for which is that using animals for food need not be incompatible with respecting important aspects of their welfare and life-style.

On this view, it seems possible for the genetic engineer to avoid the charge of disrespect if the material with which he or she is working is not seen in a purely instrumental light, but there are complex questions to be tackled here about the 'ends' of animals and plants, which we shall return to in Part 2.

Secondly, on a less abstract level, why should it be assumed that any scientist is *likely* to exhibit a lack of respect in the area of his or her particular expertise? Specialist knowledge and skills commonly lead to *greater* rather than *less* sensitivity, awareness and awe. Astronomers do not despise the heavens because they know a lot about them, nor do veterinary surgeons lose their respect for animals because they are skilled in treating them and operating upon them. So why should genetic engineers be thought automatically to lack respect for the living material which they are working with?

Thirdly, genetic engineering seems no more or less open to the charge of disrespect than traditional techniques. Selective breeding has always aimed at modifying life-forms for commercial ends and has, therefore, treated plants and animals partly at least as 'means'. Do genetically manipulated tomatoes exhibit any more or less disrespect for Nature than the amateur gardener's 'stringless' runner beans or F_1 hybrid cabbages?

Summary

In this chapter, we have started to examine some moral and ethical concerns about genetic engineering and to demonstrate their importance.[30] A possible distinction between 'moral' and 'ethical' has been suggested and questions raised about how ethical methods may be used to evaluate moral concerns. We have distinguished between extrinsic and intrinsic concerns and have investigated examples of each that are relevant to all applications of genetic engineering. Extrinsic concerns about risk and safety cannot be settled by giving clear-cut, straightforward answers, for complex judgments will always have to be made about the justifiability of the risks taken and the weighing of possible costs against possible benefits. Intrinsic concerns to the effect that genetic engineering is unnatural or shows disrespect towards Nature were shown to rest on a variety of concepts and assumptions some of which lack clarity and coherence.

In the discussion in this chapter, we have tried, where possible, to avoid reference to any religious arguments, viewpoints or considerations. Genetic engineering does, however, raise religious questions for many people, whose moral concerns will reflect this additional factor, and it is to these questions that we will turn in the next chapter.

4
Theological concerns

Genetic engineering is . . . dealing with physical processes which are
part of that divine providence and have a place in God's future
for the universe.

Jones (1991)[1]

. . . the devil is already at the door, cleverly disguised as an engineer
and entrepreneur.

Rifkin (1993)[2]

Why devote a chapter in a book on the science and ethics of genetic engineering to theological concerns? Two main answers can be given. First, surveys consistently show that the majority of people, when asked, affirm a belief in God.[3] Although a stated belief in God may not translate into any overt religious activity, such as church involvement, it often connects with what people feel about important issues in life. There is also evidence from telephone surveys and other questionnaires that a significant number of people in some countries object to genetic engineering on the grounds that it is against God's will or contrary to their religious beliefs.[4] So theological concerns about genetic engineering matter because many people have them.

Secondly, all the world's religions have teachings concerned with creation and with human nature. Because genetic engineering is sometimes considered unacceptable on the grounds that it involves scientists trying to improve on 'God's creation', it is worth looking in some detail at theological understandings about creation.[5]

A chapter on theological concerns about genetic engineering must address two major issues before concerning itself with the specifics of the subject. First, how does ethical thinking within a religious framework relate to secular moral reasoning? Are the two wholly distinct? Is dialogue, even rapprochement, between the two a possibility? Or is the most that can be hoped for a sort of mutual tolerance? Given that almost all of us live nowadays in a pluralist society, namely a society where there no longer exists a single shared set of values and beliefs, such questions are crucial. At the very least, the members of such societies benefit from understanding something of each other's viewpoints.

Secondly, how can a chapter on theological concerns avoid simply repeating the beliefs and prejudices of its authors? What claim to generality can it have? To be precise, given that the two of us both belong to the Protestant tradition within Christianity, how can we claim fairly to represent the views of other Christian denominations, not to mention the many other religions that exist? To a large extent, we can't. We belong to a particular religious tradition, just as we are both white, male, academic and middle class. To pretend we aren't would be silly, misleading and unhelpful. Inevitably, what we write will be coloured by who we are and what we think and believe. At the same time, there are certain significant similarities between religions, which we shall outline below, and it may be that a grounding in one of them aids, rather than precludes, at least a partial understanding of the ethical concerns of the others.

In this chapter, we first look at the relationship between ethical thinking within a religious framework and secular moral reasoning. We then go on to look at the nature of religious belief and examine, in particular, the claims religions make to know what is right. Having thus cleared the decks, as it were, the third part of the chapter examines religious thinking about the natural world and the relationship of humanity to the rest of the creation. Finally, we review what people of various religious faiths feel about genetic engineering.

Religious and secular ethical reasoning

What might be the relationship between ethical thinking within a religious framework and ethical thinking without a religious framework (i.e. secular)? One extreme answer that has been given is to say that all ethical reasoning depends on religious belief. For many people, such a view is immediately contradicted by the existence of people without a religious faith but with strong moral principles that translate into right actions. However, despite this objection, the thesis 'all ethical reasoning depends on religious belief' can be defended in a number of ways. For instance, it might be that a good and caring atheist is good because she inherits a set of views, for example that harming other people is wrong, from the society in which she grew up; even today, practically all societies still derive many of their values from religious ones. A consequence of this argument is that should our 'good and caring atheist' ever realise that her moral actions are merely the inherited legacy of religious values, she might discard them, to lead from then on a life of pure selfishness.

Although the argument that all ethical reasoning depends on religious belief may appeal to some, we will not rely on it for two reasons. First, to many readers it probably just doesn't sound that convincing. There are simply too many people without religious beliefs who try to do what is right, rather than act amorally. Secondly, it requires that actions and intentions are good, not in themselves, but because they are deemed good by God. For if, for example, not harming other people is good even in the absence of the existence of God, then all moral beliefs do not depend on religious beliefs. What is at issue here is the question posed by Plato in the *Euthyphro*: 'Do the gods love holiness because it is holy, or is it holy because they love it?' Doesn't it seem unacceptable to say that right is right and wrong is wrong only because God deems them so? Even if it somehow seems to limit God, there must be some sense in which, for example, justice, compassion and honesty are good, and not merely because God considers them so in arbitrary preference to injustice, selfishness and dishonesty.[6]

A different view of the relationship between secular and religious ethical reasoning can be advanced and is likely to be favoured by many religious believers. This view accepts that certain fundamental principles and behaviours, such as justice, compassion and honesty, are good irrespective of religious belief. However, it argues that there are two other categories of right actions. First there are those right actions that are right only for the adherents of the particular religion, for example the keeping of certain food laws and other religious rituals. Then there are other activities, such as sexual faithfulness within marriage, that are held by many religious believers to be right for *all* people, even though such activities may not be acknowledged as right by all non-believers.

It is this last category of right actions that is of most significance for our purposes. For example, suppose I believe that God created sheep and cattle as *discrete* kinds of animal. I may well reject a technology that involves putting cattle genes in sheep or *vice versa*, let alone putting plant genes in bacteria or even human genes in mice. Further, I may not be persuaded that such genetic engineering is acceptable *whatever* the benefits. For me, it is intrinsically wrong.

The nature of religious belief and practice

All religions lay claim to distinctive ways of knowing what is right. So here we first of all survey the common features of different religions and then examine the question 'How do we know what is right to do?' and see how it may be answered from both the secular and the religious perspective.

Hardly surprisingly, it is rather difficult to answer the question 'What is a religion?' in a way that satisfies the members of all religions. Nevertheless, the following are generally characteristic of most religions.[7]

The practical and ritual dimension

The practical and ritual dimension encompasses such elements as worship, preaching, prayer, yoga, meditation and other approaches to stilling the self.

The experiential and emotional dimension

At one pole, the experiential and emotional dimension includes the rare visions given to some of the crucial figures in a religion's history, such as that of Arjuna in the *Bhagavadgita* and the revelation to Moses at the burning bush in *Exodus*. At the other pole, are the experiences and emotions of many religious adherents, whether a once-in-a-lifetime apprehension of the transcendent or a more frequent feeling of the presence of God either in corporate worship or in the stillness of one's heart.

The narrative or mythic dimension

All religions hand down, whether orally or in writing, vital stories. Technically, these are often described as myths, for example the myth of the six day creation in the Judaeo–Christian scriptures. For some religious adherents, myths are believed literally, for others they are understood symbolically.

The doctrinal and philosophical dimension

The doctrinal and philosophical dimension arises, in part, from the narrative dimension as theologians within a religion struggle to integrate these stories into a more general view of the world. Thus, the early Christian church came to its understanding of the doctrine of the Trinity by combining the central truth of the Jewish religion – that there is but one God – with its understanding of the life and teaching of Jesus Christ and the working of the Holy Spirit.

The ethical and legal dimension

If doctrine attempts to define the beliefs of a community of believers, the ethical and legal dimension regulates how believers act. So Islam has its five Pillars – Shahada (profession of faith), Salat (worship), Zakat (alms-giving), *saum* (fasting) and Hajj (pilgrimage) – while Judaism has the ten commandments and other regulations in the Torah, and Buddhism its five precepts.

The social and institutional dimension

The social and institutional dimension of a religion relates to its corporate manifestation. For example the Sangha – the order of monks and nuns founded by the Buddha to carry on the teaching of the Dharma – in Buddhism, the umma – the whole Muslim community – in Islam, and the Church – the communion of believers comprising the body of Christ – in Christianity.

The material dimension

Finally, there is the material dimension to each religion, namely the fruits of religious belief, as shown by places of worship (e.g. synagogues, temples and churches), religious artefacts (e.g. Eastern Orthodox icons and Hindu statues) and sites of special meaning (e.g. the river Ganges, Mount Fuji and Ayer's Rock).

Secular and religious approaches to deciding what is right

How do we know what it is right to do? This enormous question, which has exercised the minds of philosophers and others since before the advent of writing, must be addressed, even if only in outline. In Chapter 3, we asked what ethical methods could be used to evaluate moral concerns and referred briefly to a wide range of competing ethical theories. We now need to look at some of these in a little more detail in order to examine the distinctive claims that religions make to knowing what is right and wrong. Let us first adopt a wholly secular perspective and suppose that there is no God, or, more formally, that even if there is, (s)he need not be assumed to exist for our argument.

Most people would argue that the right thing to do cannot depend on one's point of view. For example, we would reject the notion 'It is right for people in the country where I live to enjoy, in perpetuity, a higher standard of living than people in other countries', however attractive such a notion might be, on the grounds that it is arbitrary. Why should it matter in which country I live? Why should my individual comfort provide any sort of ethical criterion?

As a result of this line of argument, most theories of ethics end up treating people, at least in some sense, as equals. To this family of argument belongs Immanuel Kant's categorical imperative that moral principles must be universal and not dependent on an individual's wishes, inclinations and idiosyncracies, and its parallel formulation that people must be treated as ends, not means. A vital question that arises from this form of reasoning is 'Who are persons'? Are severely disabled babies persons? What about all human embryos? Are persons restricted to the species *Homo sapiens*? What about the great apes and other sentient animals? We shall consider these questions in more detail in Chapter 7 in the context of the genetic engineering of non-human animals for human ends.

The ethical theories of Kant and his followers are often labelled deontological because they emphasise a duty-based morality that makes certain acts, prescribed by universal principles, obligatory, regardless of their consequences. The intrinsic concerns about genetic engineering described in Chapter 3 fall into this category. By contrast, utilitarianism, which was mentioned in connection with the extrinsic concerns examined in Chapter 3, argues that consequences are all-important and that actions can be ethically evaluated only in terms of the overall benefit produced.

Then there is the apparently rather modest approach of non-maleficence. This stems from the maxim *Primum non nocere*, 'First, do no harm', contained in the Hippocratic oath. Although at first it may appear subordinate to utilitarianism, it is easy to imagine circumstances under which it would take precedence. Suppose, for example, that a cure for breast cancer was discovered that could only operate by a certain chemical being introduced into the water system. Suppose further that the consequences of this treatment were to save the lives each year of 10 000 people with breast cancer *and* to cause the deaths of 1000 people who did not have breast cancer. Although a utilitarian approach might argue for the intro-duction of the chemical into the water system, the principle of nonmaleficence would argue against it.[8] However, as we showed in Chapter 3 when looking at the possible consequences of genetic engineering, it is often difficult if not impossible to determine which option is likely to do least harm.

So there are various secular ways in which general principles can be generated in attempting to answer the question 'What is it right to do?'. Some of these principles claim that certain actions are right or wrong in themselves while others maintain that whether a particular action is right or wrong depends on its consequences. The question we can now address is 'What distinctive claims do religions make to knowing what is right and wrong?'.

What distinctive claims do religions make to knowing what is right and wrong?

The short answer is 'several'. It is apparent from our earlier analysis of the seven dimensions characteristic of religions that religions have a number of reasons for possessing a distinctive understanding of right and wrong. For a start, most religions have sacred scriptures believed to contain teachings either of God or of pivotal early leaders of the faith. Then, religions have teachings and traditions assembled over many years (e.g. those contained in the Jewish Talmud and the teachings of the Roman Catholic Magisterium). Finally, most religions maintain that our conscience, namely what each of us believes to be right and wrong, has some divine justification.

For all these reasons, religions contain distinctive teachings about the nature of right and wrong.[9] But before going on to look at religious understandings of creation and human nature, it will help if we look at the issue of natural law, for here is a phenomenon that both secular and religious writers have adduced to support their arguments, but in different ways.

Natural law

The simplest understanding of natural law is that by virtue of human reason we can discern something of how things ought to be by looking around us (i.e. at nature) and seeing how things are.[10] For instance, to give a trivial example, we know that it is natural for parents to look after their children – children beneath a certain age are not yet capable of complete independence nor are they fully responsible for their actions. Accordingly, we instinctively feel that it is wrong for young children to be left alone and unattended for long periods of time. We could discuss *ad nauseam* for how long children of a particular age might rightly be left unattended, but the principle is clear.

However, it can be objected that the theory of natural law immediately runs into all sorts of problems, as outlined in Chapter 3. We might agree that it is unnatural, and hence wrong, for parents to leave young children alone for long periods of time, but what about keeping animals in cages? In the wild, mammals run around, play and have lots of space, while birds have the freedom to fly. Manifestly, then, it is unnatural for countless pigs, calves and chickens to spend their entire lives indoors in factory farms deprived of natural light and the freedom to move, let alone play. But is factory farming wrong? Some would argue yes, it is wrong, on the grounds that it is unnatural, quite aside from the issue of whether or not the animals suffer. Others would still accept the force of natural law but argue that it is natural for farmers to domesticate animals, as they have for countless thousands of years.

So even if we accept the principle of natural law, it can be difficult to discern what is natural and what is unnatural. Is it natural to relieve suffering through medicine? Are contraception and abortion natural? Are differences between male and female behaviour natural?

As we saw in Chapter 3, a more general problem with natural law can be raised, namely that its supporters lay themselves open to the objection that they commit the naturalistic fallacy that assumes that one can deduce what ought to be from what is. For example, to deduce from the observation that parents smack their children the principle that parents should smack their children, or, more extremely, to deduce from the observation that women are generally lower paid than men the principle that women *should* be lower paid than men.

This objection to natural law can be countered, at least to some extent, from a theological viewpoint as follows. The world (i.e. 'nature' or 'what is') is the work of God and, therefore, tells us something about God. As God is good, we can deduce what is good from the world. A more refined version of this argument accepts that nature, as we see it today, is corrupted or imperfect or fallen, but that sufficient remains of the original creation for us, at least sometimes, to gain insights as to how things ought to be. The first three chapters of *Genesis*, for example, describe a world in which people lived for ever, where food was freely available and could be obtained with no hardship, where childbirth was painless and safe, where men and women lived in mutual harmony, and where everyone was vegetarian.

Religious understanding of the creation

In our consideration of natural law, we have already touched upon religious understandings of the creation, seeing how many religions view the original creation as perfect but recognise that, in its present state, it is far from that. It should be noted that the phrase 'the creation' is used here in a wider sense than in some secular cosmologies, to mean both 'the original creation' and 'that which exists today'.

The question we need to consider is 'What is to be *our* role in creation?' Manifestly genetic engineering involves humans creating new types of organism. To decide if this is blasphemous, wrong for a different religious reason or justifiable from a religious perspective requires us to adopt some theological view about how humans should interact with the rest of the created order (i.e. the natural world). We shall now consider the variety of views held by the various religions on this issue. Of course, any one religion contains within itself a range of views almost as great as that between very different religions. Our analysis may, therefore, be felt by some to generalise and oversimplify, and by others to contain too many caveats and qualifications. Our purpose is not so much to analyse in any depth differences between religions but to show the variety of approaches that religions have to this question.

Leave well alone

The simplest theological response to the question of how humans should relate to the rest of creation is that we should leave well alone: 'When you start playing around with genes, you're playing God. I don't think we have enough experience to play God. We need a little humility. It seems to be in short supply these days.'[11] The difficulty with this argument is that it invites us to raise the question 'What's so special about genetic engineering?' For the same argument could be raised against *any* new technology. As we discussed in Chapter 3, scientists and technologists can never guarantee that a new process or product is totally safe. This is not to dismiss the suggestion that 'we need a little humility'. On the contrary, the advent of a genuinely new technology – and genetic engineering is certainly that – suggests that we should proceed with caution. But this is an argument about safety and about the pace of change, not specifically about whether genetic engineering is right or wrong.

Of course, specific religious objections to genetic engineering can be raised. Indeed the phrase 'playing God' in the above quotation provides one of them, and such fears about human 'hubris' or arrogance in trespassing upon divine territory go back at least as far as the religious beliefs of the Ancient Greeks. These objections will not be answered by extrinsic considerations such as assurances of safety or promises about the ultimate benefits of genetic engineering to us all. We can note that similar objections were raised in the nineteenth century when the first anaesthetics were developed and, of course, some religious believers, notably Christian Scientists, refuse medical help to this day. Nevertheless, most other religious believers have reached different views and nowadays accept the benefits of modern medicine, combining a trust in medicine with their religious faith and seeing medicine as a gift of God.

Indeed, if 'playing God' includes breeding plants that could never have arisen in the wild, then almost everyone, with the exception of some aboriginal tribespeople, either plays God or benefits from someone who has. All of the world's crop plants, to say nothing of domesticated animals, are the products of thousands of years of artificial selection. It is inconceivable that the human race could go back to a pre-agricultural existence. Anyway, 'playing God' cannot be restricted to the domestication of animals and plants. Presumably it must also logically include a tremendous range of human achievements from the synthesis of medical drugs (a valuable supplement, most would argue, to those obtained directly from wild plants) to the invention of electricity (rather than relying solely on the Sun for light).

On the assumption that humans will continue, quite rightly, to benefit from such activities as agriculture, medicine and industry, what other approaches are there to how we should relate to the rest of creation? The most straightforward is that we should exploit the natural world for our own ends.

Exploitation for human ends

Traditionally, a strong strand within certain religions, particularly Judaism, Christianity and Islam, has viewed the created order as having been created for humans to exploit. In *Genesis*, Adam and Eve are told, 'Be fruitful and increase, fill the earth and subdue it, rule over the fish in the sea, the birds of heaven, and every living thing that moves upon the earth'.[12] It is difficult for us to reconstruct with certainty how verses such as this one were first understood. There is little doubt, however, that the understanding that humans were created in 'the image of God'[13] led many people to feel that they were set over nature and had authority to do with it pretty much as they liked. Christianity, in particular, has often been accused of a rapacious attitude towards nature: 'Especially in its Western form, Christianity is the most anthropocentric religion the world has seen. In absolute contrast to ancient paganism and Asia's religions, it not only established a dualism of man and nature but also insisted that it is God's will that man exploit nature for his proper ends'.[14]

Certainly, the expectation, particularly within some Christian traditions, of life after death, an imminent Apocalypse and the advent of 'a new heaven and a new earth'[15] made it easy for people to feel that they could ignore the needs of the rest of the created order and simply use this world for their ends.

Stewardship/mutuality

The notion that humans can exploit the whole of the rest of the created order for their own ends is relatively uncommon in religions. More commonly, religions agree that humans have, at the very least, a responsibility as to how they use the creation.

In Hinduism, the world's oldest major religion, all life is sacred. Visnu, as supreme being, endlessly creates the worlds of matter and withdraws it into his existence time after time as the cycle of seasons endlessly repeats itself.[16] In the Vedic literatures, mother Earth is personified as the goddess Bhumi, or Prithvi, the abundant mother who showers her mercy on her children. It is not surprising that Hinduism views humanity as having a great responsibility towards the Earth:

> According to the Isa Upanishad, this planet does not belong
> to humanity, any more than it belongs to the other species
> living on it: 'Everything in the universe belongs to the Lord.
> You should therefore only take what is really necessary for
> yourself, which is set aside for you. You should not take
> anything else, because you know to whom it belongs.' So
> long as we treat the planet carefully and take only our share,
> acknowledging that it and everything else belongs to God, the
> planet will provide for our needs; but as soon as we try to take
> nature's gifts without offering anything in return we become
> no better than thieves.[17]

In Judaism too, there is a strong emphasis on the responsibilities that humans have towards nature. Agricultural land was supposed to lie fallow every seventh year as a 'sabbath of sacred rest'.[18] Further, every 50 years, on the Day of Atonement in the Jubilee Year, all land must return to its original owner. Because the Earth is the Lord's, no one has unconditional land rights.[19] The Jewish scriptures also include a number of instructions relating to animal welfare, while some of the Wisdom writings argue that creation has a purpose beyond that of human benefit. For example, in the book of Job, the Lord answers Job out of the tempest:[20]

Who has let the Syrian wild ass range at will
and given the Arabian wild ass its freedom?
I have made its haunts in the wilderness
and its home in the saltings;
it disdains the noise of the city
and does not obey a driver's shout;
it roams the hills as its pasture
in search of a morsel of green.
Is the wild ox willing to serve you or spend the night in your stall?
Can you harness its strength with ropes;
will it harrow furrows after you?

It should not be thought that the understanding in Judaism of the need for human stewardship is entirely a modern phenomenon. In the twelfth century of the common era, the Jewish scholar Abraham ibn Ezra said of Psalm 115.16 ('The heavens are the heavens of the Lord, and he gave the earth to people'):

The ignorant have compared man's rule over the earth with God's rule over the heavens. This is not right, for God rules over everything. The meaning of 'he gave it to the people' is that man is God's steward (*paqid* – officer or official with special responsibility for a specific task) over the earth, and must do everything according to God's word.'[21]

In Buddhism there is a very strong emphasis on how we should relate to the natural world; for example, there is a prohibition on the taking of animal life. Although Buddhism exists in many different forms, human responsibility towards the creation is a common theme, though the word 'creation' is somewhat inappropriate as the Buddha taught that there is no creator God as the first cause, because there is no beginning. While Buddhism teaches that humans, unlike other creatures, have the opportunity to realise enlightenment, it does not teach that humanity is superior to the rest of the natural world. Indeed, the doctrine of 'emptiness' in Buddhism, as originally developed by the philosopher Nagarjuna, in asserting that all things are empty simply denies that anything can exist on its own.[22]

The Chinese Buddhists, in particular, emphasised the intimate connections between all things. This philosophy found a more concrete expression in the Zen tradition. For instance, the Japanese Zen master Dogon says:

> It is not only that there is water in the world, but there is a world in water. It is not just water. There is also a world of living things in clouds. There is a world of living things in the air. There is a world of living things in fire. There is a world of living things on earth. There is a world of living things in the phenomenal world. There is a world of living things in a blade of grass. There is a world of living things in one staff. Whenever there is a world of living things, there is a world of Buddha ancestors. You should examine the meaning of this.[23]

In related vein, the contemporary Vietnamese monk and poet Thich Nhat Hanh writes:

> When we look at a chair, we see the wood, but we fail to observe the tree, the forest, the carpenter, or our own mind. When we meditate on it, we can see the entire universe in all its interwoven and interdependent relations in the chair. The presence of the wood reveals the presence of the tree. The presence of the leaf reveals the presence of the sun. The presence of the apple blossoms reveals the presence of the apple. Meditators can see the one in many, and the many in one. Even before they see the chair, they can see its presence in the heart of living reality. The chair is not separate. It exists only in its interdependent relations with everything else in the universe. It *is* because all other things *are*. If it is *not*, then all others are *not* either.[24]

The notion of stewardship is also a significant theme in much of the writing in recent decades on Christianity,[25] Islam[26] and other religions.[27] Narrowly understood, stewardship, as a view of how humans should relate to the rest of the natural world, still implies something of a 'them' (the rest of the natural world) versus 'us' (humans) situation. However, there is a more holistic view of our relationship to the rest of creation. This relationship can be described as one of mutuality, reflecting the biological understanding that a relationship is mutual if it benefits both partners. The word 'mutuality' is, therefore, intended to avoid overtones of hierarchy and superiority. As some of the above quotations, for example that from Thich Nhat Hanh, have indicated, humanity can be viewed as having a depth of inter-relatedness with the rest of the natural order that transcends an understanding which has humanity somehow set apart from, or above, the rest of creation.

George Herbert (1593–1633), the Anglican priest and poet, saw the relationship of humanity to the rest of the created order as one of a priest to his flock:

> Man is the worlds High Priest: he doth present
> The sacrifice for all: while they below
> Unto the service mutter an assent (*Providence*)

In like vein, Arthur Peacocke has explored related metaphors for our relationship to the rest of creation, including humanity as prophets of creation, lovers of creation or fellow-sufferers with it.[28]

Co-creation

Another interpretation of how humans relate to the rest of the natural world is that, in some sense, we are co-creators, co-workers or co-explorers with God.[28,29] Although this may at first sound blasphemous, or simply downright silly, to some readers, the reasoning goes as follows. Our scientific understanding of the universe, in particular cosmology and the biological theory of evolution, shows that creation is an ongoing process. The universe has been in a continual state of development for some fifteen thousand million years. Within just the last few thousand years, humans have begun consciously to influence the course of that continued creation in a way never before attained by any species. In any useful sense of the term, therefore, we are co-creators with God, altering the future of the world year by year, hour by hour, as we cause some species to become extinct and alter the genetic constitution of others by the traditional techniques of artificial selection as well as by the newer approach of genetic engineering. Arthur Peacocke puts the point more poetically:

It is as if man has the possibility of acting as a participant in creation, as it were the leader of the orchestra of creation in the performance which is God's continuing composition. In other words man now has, at his present stage of intellectual, cultural, and social evolution, the opportunity of consciously becoming *co-creator* and *co-worker* with God in his work on Earth, and perhaps even a little beyond Earth. To ask how to fulfil this role without the *hubris* that entails the downfall classically brought upon those who 'would be as gods' is but to pose in dramatic fashion the whole ecological problem. But at least one who sees his role as that of *co*-creator and *co*-worker with God might have a reasonable hope of avoiding this nemesis, by virtue of his recognition of his role *ipso facto* as auxiliary and co-operative rather than as dominating and exploitative.[30]

Some have argued that the term co-creator overstates the human role. Unlike God, we are not creators out of nothing, nor did we create the universe; we are only creators in a very subsidiary sense, and that on one small planet.[31]

Religious approaches to genetic engineering

Genetic engineering is increasingly being examined from a theological perspective, though most of what has been written is the product of Christian writers. A helpful summary of the range of views on genetic engineering found among the major world faiths is provided by J. Robert Nelson.[32] A number of specific theological analyses of genetic engineering will be addressed in later chapters, but here we can point out three main approaches.

Rejection

Some authors reject genetic engineering on the grounds that genetic engineering entails humans having too much power over animals. Andrew Linzey, for example, considers genetic engineering to be a form of slavery:

> . . . genetic engineering represents the concretisation of the
> *absolute* claim that animals belong to us and exist for us. We
> have always used animals, of course, either for food, fashion
> or sport. It is not new that we are now using animals for
> farming, even in especially cruel ways. *What is new is that we
> are now employing the technological means of absolutely subjugating
> the nature of animals so that they become totally and completely
> human property.*[33]

We will explore in subsequent chapters the ethical issues surrounding the ownership and patenting of life. Here we can note
that certain animals, not to mention plants, have been considered
human property in practically every society since time immemorial.
Farm animals and pets, in particular, are always owned by people.
Indeed, traditional techniques of animal breeding have already
rendered some species incapable of surviving without human
assistance: both domesticated turkeys and domesticated camels,
for example, are unable to breed without human aid. Whether this
is right or not needs to be discussed from a wider perspective than
that concentrating on theological concerns. We will look in
Chapter 7 at the twin issues of human responsibilities towards
animals and whether animals have rights.

Another possible reason for rejecting genetic engineering is that
it involves too exploitative a view, not just of animals but of all of
nature. This point is made by K. S. Satagopan:

> Nature has its own genetic techniques which is borne by the
> change from Amoeba to men. But man's attempt to speed up
> change or to bring about important changes in specific ways
> and forms is very different. It is violence since it is against
> ecology. All Indian religions (Jainism, Buddhism and
> Hinduism) are against violence, implicitly or explicitly. From a
> theoretical point of view Indian religions cannot have anything
> favourable to say about genetic engineering.[34]

This view echoes Martin Heidegger's argument that in technology we make objects according to some blueprint that we determine. We design things to satisfy our purposes rather than allow
our purposes to be affected by, and find creative expression

through, the qualities of the objects themselves.[35] Heidegger's point is even more salutary when applied to genetically engineered organisms.

As the interviews and questionnaires referred to earlier showed,[4] many religious believers reject genetic engineering not because of concerns specific to the moral status of animals but simply because it clashes with their understanding of God's action in the world. Arguments from the scriptures may be adduced to support a person's beliefs. Though we have no evidence to support the assertion, it seems possible that the more fundamentalist and less liberal a person's religious faith, the more likely they are to reject genetic engineering, unless they perceive very considerable benefits from it. For instance, someone who believes in the literal creation of the universe in six days is presumably less likely to support genetic engineering than someone, whether religious, agnostic or atheistic, who accepts the theory of evolution.

Caution

The most frequent response by religious writers to the issue of genetic engineering is one of caution or hesitancy. Darryl Macer, for example, writes that: 'There needs to be a balance between our creativity and caution'.[36] Similarly, Caroline Berry writes:

> The issues facing Christians are related . . . to the tensions
> that arise as we discover more of our genetic make-up and
> apply that to the workaday world:

- the tension between seeing a person as the sum of their genes and as being someone made in the image of God and for whom Christ died;

- the tension between wanting the best for our children and accepting them as they are, with their abilities or disabilities, whether mild or severe – we must continue to see children as gifts from God rather than assets to be acquired;

- the tension between the individual and their right to privacy and autonomy and the needs of society as a whole;

- tensions between commercial profits, the expense of research and our
  knowledge that a large proportion of the world's population are in need
  of food and clean water.'[33]

In particular, some people may hesitate about the movement of
genes between humans and other species, fearing that this somehow diminishes the distinctiveness of being human. For example,
the notion that humans are made *imago Dei* may cause some with a
Christian faith to feel uncomfortable about a technology that may
threaten to blur the dividing line between humans and the rest of
the created order. Of course, others may feel differently, perhaps
believing that all of creation is, in a way, *imago Dei*, in the sense that
how can that which is created do other than reflect its creator,
sustainer and redeemer.

Acceptance with caveats

Finally, though this category overlaps with the previous one, there
are religious writers who accept genetic engineering, though
typically with certain specific caveats. Phil Challis, arguing against
Andrew Linzey's rejection of genetic engineering considered
above, writes:

> We are co-creators with God, "fearfully and wonderfully
> made" (Ps139:14). With our finite freedom we are called by
> Him to act responsibly as we continue the process of genetic
> manipulation of domestic organisms. A theology that
> emphasises embodiment rather than body–spirit dichotomy,
> that emphasises becoming rather than immutability as an
> essential part of God's nature, that emphasises relationship
> within the web rather than domination from outside the
> system, such a christian theology may provide a critical
> framework that can realistically embrace the potential of
> genetic engineering for good.[38]

Ronald Cole-Turner[39] has explored the implications of a distinction between humans as co-creators with God – a concept
which, he feels, contains a number of difficulties – and humans as
participants, through genetic engineering, in redemption. Here

redemption is being used in the sense of 'restoration'. The idea is that genetic engineering can help to overcome genetic defects caused by harmful mutations. In this way, genetic engineering can help to restore creation to a fuller, richer existence and can, Cole-Turner maintains, play an important role without encroaching on the scope of divine activity.

It can perhaps be argued, in this vein, that humans may have a theological responsibility, even a duty, to use genetic engineering to root out imperfections in the natural world, including those found in humans. Viewed in this light, genetic engineering can be seen as a tool with the potential to eliminate harmful genetic mutations, reduce suffering and restore creation to its full glory.

Conclusions

To some extent the very writing of a chapter specifically on theological concerns admits to a dichotomy of secular and religious thinking. Such a division would be rejected by many, certainly by most religious believers. At the very least, a theological view of genetic engineering rejects a reductionist, materialistic understanding of the natural world, and indeed such a view is also rejected by some secular commentators.[40] It is clear that, while the theology of genetic engineering is still in its infancy, there exists a range of views. Perhaps the most widespread view is one of caution about the rightness of genetic engineering, though some religious writers are prepared to condemn it entirely.

In our view, it is difficult to maintain fundamental theological objections to all aspects of genetic engineering *per se*. The notion that in some sense to be human, some would say to exist in the image of God, is to be called to participate responsibly in the ongoing work of creation, is a persuasive one, though not to be undertaken lightly. The approach we will take in Chapters 5 to 8 is to proceed on a case-by-case basis, subjecting particular case studies to ethical scrutiny. Where appropriate, specifically theological issues will be raised and addressed.

PART 2

Overview

Part 2 consists of four chapters: Chapter 5 on the genetic engineering of microorganisms, Chapter 6 on the genetic engineering of plants, Chapter 7 on the genetic engineering of animals other than ourselves, and Chapter 8 on the genetic engineering of humans. In each of these chapters, we first survey the range of applications and possible implications of genetic engineering and then examine some case studies in more depth. The science behind each case study is described, and the implications are analysed in terms of the ethical principles outlined in Chapters 3 and 4.

Before getting on to the four chapters in Part 2, a word of clarification is needed about the way in which case studies are allocated to these chapters. We have chosen to focus on the organism that is genetically engineered. For example, in the production of human insulin, it is microorganisms (bacteria or yeast) that are genetically engineered, not humans. This is because the human genetic material is unchanged, whilst the genetic material of the microorganism used is altered through the insertion of the gene for human insulin into its make-up. For this reason, human insulin is examined in Chapter 5 on the genetic engineering of microorganisms, rather than in Chapter 8 on the genetic engineering of humans.

There is something of an ascending scale as we go from Chapter 5 through to Chapter 8, in that while microorganisms are genetically engineered without plants, non-human animals or humans necessarily being genetically engineered, the reverse is less often the case. The genetic engineering of plants, non-human animals and humans often – though not always – involves the genetic engineering of microorganisms. For this reason, it makes sense to deal with genetically engineered microorganisms first of all.

There is a second sense in which there is something of an ascending scale as we go from Chapter 5 through to Chapter 8. This is because the genetic engineering of humans poses ethical questions that most people feel are not posed by the genetic engineering of microorganisms, plants or non-human animals. Similarly, the genetic engineering of non-human animals poses ethical questions in addition to those raised by the genetic engineering of microorganisms and plants. For this reason, too, it makes sense to deal with genetically engineered microorganisms first, then with plants, then with non-human animals and then with humans. However, we don't want to overemphasise the importance of this order. It is useful rather than essential for our argument.

5

The genetic engineering of microorganisms

*Early in 1988 Monsanto participated in a series of focus groups, or
neighbourhood workshops, organised to exchange views on BST
[genetically engineered bovine somatotrophin] with farmers,
veterinarians, doctors, teachers, nurses and consumers. At the end of
one such meeting a local National Farmers' Union (NFU) executive,
who had organised that meeting, presented two glasses of milk to two
women, representatives of the Townswomen's Guild. He explained that
one contained milk from a BST-supplemented cow and the other
contained 'normal' milk. He asked them to choose which they
preferred. Without hesitation both responded, 'We cannot choose, there
is no choice, they're both the same.'*

Deakin (1990)[1]

*Genetically altered micro-organisms pose risks to human health and to
the environment. Scientists and regulators promise to keep that risk
to a minimum, but our own history warns of the 'whoops' theory of
risk assessment (something can go wrong – or perhaps already
has – and those affected are left with the apology, 'Whoops, sorry,
we made a mistake').*

Spallone (1992)[2]

Introduction

Microorganisms are, literally, very small organisms. They include
bacteria, viruses, yeasts, single-celled algae and single-celled
protozoa such as amoebae. To the naked eye they are invisible.

Many of them are of great benefit to humans. For example, if it wasn't for microorganisms, dead animals and plants would pile up. Instead, they decay and their nutrients are returned to the environment to the benefit of other organisms. Other microorganisms harm us. Most diseases are caused by microorganisms. For example, viruses are responsible for colds, flu, mumps and AIDS; bacteria cause cholera, diphtheria and tetanus; while protozoa cause malaria and sleeping sickness.

As we saw in Chapter 1, humans have long since used microorganisms for such processes as bread and alcohol production. Since the 1970s, scientists have learned how to alter the genetic make-up of certain microorganisms through genetic engineering. Genetically engineered microorganisms are already being used for a whole range of purposes. They can be used to synthesise complicated molecules, such as insulin and growth hormone, missing in certain humans; they can be used to manufacture vaccines, antibodies, food additives, bread, cheese and the active component of biological washing powders; they are being incorporated into biosensors able to detect minute amounts of substances for diagnostic purposes; they can even prevent water from freezing.

Looking to the future, the use of genetically engineered microorganisms is likely to increase even further. For instance, designer bacteria are being devised to help clear up oil spills and extract valuable minerals from low-grade ores. In this chapter, though, we concentrate on present practice rather than looking to possible future developments. Five contrasting examples of the use of genetically engineered microorganisms are reviewed. It might be thought that the genetic engineering of bacteria, viruses and yeasts would raise few ethical issues. However, as we shall see, this is not in fact the case.

Human insulin

Worldwide, approximately fifty million people have diabetes (or diabetes mellitus to give it its full name, which distinguishes it from a completely unrelated disease called diabetes insipidus). Diabetes occurs when the body has problems controlling the level of sugar in the blood. Having a relatively constant blood sugar level is very important and serious medical consequences can result if levels are too high or too low.

Two distinct types of diabetes are known. The more common is called non-insulin-dependent diabetes. This type used to be known as mature-onset diabetes because it mostly occurs in people over the age of 40 and is increasingly likely the older a person is. In Western countries, such as the UK and USA, 90% of diabetes cases are of this type. The most important factors leading to non-insulin-dependent diabetes are being overweight and having a familial predisposition – in other words, having a close relative who also suffers from it.

From our point of view, though, it is the less common, though still widespread, form of diabetes that is more significant. This type is known as insulin-dependent diabetes. It used to be known as juvenile-onset diabetes as it usually develops before the person is 15 years old. As its name suggests, insulin-dependent diabetes is associated with an abnormality in the body's production of insulin.

Insulin is a hormone made by specialised cells, called β cells, in parts of the pancreas known as the islets of Langerhans. The pancreas is an organ roughly the size and shape of a lamb chop. It is situated near the liver and has a number of functions, one of which is to control the level of blood sugar, glucose, in the blood.

Suppose you eat a meal rich in carbohydrates such as starch. Within a short time, the enzymes in your gut begin to break the starch down into the glucose molecules of which it is made. The glucose molecules enter the bloodstream, causing blood sugar levels to rise. This increase in the concentration of glucose in the blood is detected by the pancreas, which – in people who don't suffer from insulin-dependent diabetes – responds by secreting more insulin. This insulin is carried round the body in the bloodstream and has several effects. For example, insulin causes the liver to absorb glucose from the blood and convert it to another carbohydrate, glycogen, which is then stored ready for when the body needs more glucose.

The net result of all this is that insulin secretion soon causes the level of glucose in the blood to fall back to its usual level. Similarly, if a long period of time has elapsed since you last ate a meal, as a result of which your blood sugar levels begin to drop, the pancreas responds by producing less insulin. A number of physiological mechanisms then come into play, the net result of which is that blood sugar levels are again maintained at close to their usual value.

However, someone who suffers from insulin-dependent diabetes produces little or no insulin. As a result their blood sugar levels fluctuate greatly, often reaching extremely high levels. This leads to all sorts of medical complications. If untreated, insulin-dependent diabetes is fatal. Fortunately, the work of the Canadian physiologists Frederick Banting and Charles Best in the 1920s led to the discovery of the role of insulin in the prevention of diabetes. Soon after, doctors began to treat diabetics by giving them injections of insulin.

Treating insulin-dependent diabetes before the advent of genetic engineering

Insulin-dependent diabetes can only be treated through the injection, up to four times a day, of insulin. Until the advent of genetic engineering, this insulin came from two sources: cattle or pigs. Cattle naturally produce bovine insulin; pigs, porcine insulin. Neither of these hormones is identical to human insulin, though all three are proteins and are made up of 51 amino acid units – amino acids being the building blocks of proteins. Bovine insulin differs from human insulin by three amino acid units; porcine insulin from human insulin by just one amino acid unit.

Bovine and porcine insulin are obtained from the pancreata of cattle and pigs slaughtered for food. It has long been realised that, although bovine and porcine insulin have allowed literally millions of people to lead relatively healthy lives, there are problems with their use. For one thing, the very fact that they differ chemically from human insulin, albeit only slightly, means that some diabetics develop reactions to them. A second problem is that if the animals from which the insulin is obtained are contaminated, for example with certain viruses, the contaminants may be passed on to people injecting themselves with the animal-derived insulin. In addition, some people have ethical objections to the use of insulin obtained from cattle or pigs.

The problem of human insulin differing chemically from bovine and porcine insulin was solved before the advent of genetic engineering through the use of certain enzymes, such as trypsin. These enzymes can alter the structure of porcine insulin so that it becomes chemically identical with human insulin: the enzymes remove the one amino acid found in pigs but not humans (alanine at position 30 of the B-chain) and replace it with its human equivalent (the amino acid threonine). The resulting insulin is known as 'human insulin emp' – emp standing for enzyme-modified porcine.

The impact of genetic engineering

Despite the manufacture of human insulin emp, there was great excitement when, in 1980, scientists succeeded in synthesising human insulin through genetic engineering. At present a number of different approaches are in use, some using bacteria, some yeast. What they all have in common is that the insulin they produce is chemically identical to human insulin and is produced by recombinant DNA technology (genetic engineering) via the methods outlined in Chapter 2. Within a very short space of time this so-called 'human insulin' was being used therapeutically. The universal assumption was that it would soon replace bovine and porcine insulin.

Indeed, within just a few years human insulin became the most frequently administered form. However, reports began to appear in the medical literature of problems associated with its use.[3] Some diabetics found that, after switching to human insulin, the symptoms they experienced when their blood levels became too low (a condition known as hypoglycaemia and commonly described as 'having a hypo') changed: 'Instead of having their usual symptoms of sweating, trembling, hunger, weakness, anxiety and palpitations they find their symptoms have changed to lack of concentration, confusion, and personality change, and it takes some while to learn to recognise the new symptoms. Others find that they have no or very little warning of hypos or that they happen so quickly that they do not have time to take action.'[4]

The significance of these changed symptoms is still controversial. Stories abound in the diabetic community of problems that have resulted, and some people are said to have died. However, some medical experts and manufacturers have denied the existence of increased hypoglycaemic unawareness with human insulin. It should be realised that the reported loss of hypoglycaemic awareness is not specifically related to the production of human insulin through genetic engineering; the same problems have been reported in diabetics using human insulin emp. In other words, the problem, presuming it exists, is related to the *switch* from one chemical form of insulin (porcine or bovine) to another (human insulin), irrespective of the way in which the latter is synthesised.

Future prospects for the treatment of insulin-dependent diabetes

Regular insulin injections – whether of bovine, porcine or human insulin – have saved the lives of millions of people. Yet they are not a perfect answer to insulin-dependent diabetes. In a person without diabetes, the body constantly measures blood sugar levels, continuously altering the pancreatic release of insulin accordingly. Injecting oneself with insulin a few times a day is a very crude mimic of this natural process. Pharmaceutical companies have responded to this problem by producing distinct formulations of insulin. Some of these act quickly, others over a number of hours. However, a recent review concludes that: 'Despite many decades of experience and improvements in the quality of insulin and the delivery devices and formulations available, serious disadvantages to conventional insulin therapy remain'.[5] These disadvantages can have serious medical consequences. For instance, long-term side effects of excessive variation in blood sugar levels include blood vessel proliferation in the eyes, which can result in significant deterioration in eyesight and even blindness. Kidney damage is also frequent.

There are at least three ways in which the problems associated with the irregular supply of insulin to the bloodstream in diabetics may eventually be solved, though each is at present at the research stage. The simplest, and most thoroughly investigated to date, is that diabetics might carry around with them a little pack that would secrete insulin continuously over a period of 24 hours. These sorts of packs are used in hospitals quite extensively for the injection of a number of different substances. It is possible that the same pack could monitor blood sugar levels, altering the rate at which it injects insulin accordingly. This would effectively mimic the precise, regulatory action of the normal control of insulin secretion.

A second possibility is that a diabetic might receive a transplant of healthy β cells – the cells that secrete insulin. Until recently the major problem with this approach, as with most transplants, has been the issue of rejection. Foreign β cells are detected as such by the body's defence system, then assumed to be germs and destroyed. Work being led by Patrick Soon-Shiong of the Wadsworth Medical Center in Los Angeles has gone a long way to overcoming this problem. Soon-Shiong effectively hides the β cells by placing them in tiny capsules made of a seaweed extract called alginate. This lets small molecules such as glucose and insulin in or out, but keeps out those components of the body's defence system – antibodies and white blood cells – that destroy foreign substances.[6]

A third possibility is that scientists may be able genetically to engineer new forms of insulin with new properties. These molecules are called 'insulin analogues'. Their structures are very similar to that of human insulin or human proinsulin (the molecule from which the body makes insulin). However, these insulin analogues contain one or more differences in their amino acid composition. The consequences can be startling.[5] Some insulin analogues last up to 10 times longer than human insulin; others, at least in the laboratory, have 10 times the potency of human insulin. This work opens up the possibility that diabetes may one day be managed through the occasional injection of a cocktail of different insulin analogues, each with its own altered pharmaceutical properties.

Preventing diabetes

So far we have taken the high incidence of diabetes for granted. It has been estimated that as much as 95% of all non-insulin-dependent diabetes is preventable through exercise and diet. However, insulin-dependent diabetes cannot, as yet, be prevented. Indeed, its precise cause or causes are still uncertain, though considerable advances in this field have been made in recent years. There are two main avenues of enquiry. The first is that insulin-dependent diabetes is triggered by a virus infection early in life. The second is that the loss of the β cells of the pancreas is the result of an auto-immune disease. In such a disease something goes wrong with the body's defence system and a part of the body, in this case the β cells of the pancreas, is mistakenly recognised as foreign and destroyed.

It is possible that if these lines of research prove fruitful we may one day be able to look towards preventing, rather than merely treating, diabetes.

Human growth hormone

Throughout our lives, our bodies produce human growth hormone, also known as human somatotrophin. This hormone, like human insulin, is a protein. It is produced in a small region at the base of the brain called the pituitary gland. From here it passes into the bloodstream and is carried round the body. Its main effects are on bones and muscles: growth hormone stimulates cells to increase in size or divide.

Some children produce too much growth hormone. As you might expect, they grow much taller than average – by the age of 21 a man whose pituitary has secreted too much growth hormone may be as much as 8 foot (2.4 metres) tall, a woman may be 7 foot (2.1 metres) tall. Other children produce too little growth hormone. They end up much shorter than average, typically around 4 foot (1.2 metres) in height. This condition is sometimes referred to as pituitary dwarfism, though the term is often avoided on the grounds that some people find the word 'dwarfism' insulting.

In most cases, people with an abnormally high or low production of growth hormone end up physically quite healthy, with a body that is normally proportioned, albeit unusually large or small. Until the 1950s there wasn't anything that could be done to change their height – eating more, for example, has no effect. Then Dr Maurice Raben at the Tufts New England Medical Center began painstakingly to extract human growth hormone from the pituitary glands of corpses. The hope was that if this was given to people with abnormally low levels of growth hormone production, they might benefit by growing more.

One of the first people Maurice Raben treated was a Canadian called Frank Hooey. By the time he saw Maurice Raben, Frank Hooey was aged 17 and was only 4 foot 3 inches (1.28 metres) in height – the height of a typical eight and a half year-old.[7] Over the next five years, Frank Hooey received thrice-weekly injections of human growth hormone. By the end of the treatment he stood at 5 foot 6 inches (1.65 metres), slightly below average, but well within the normal range.

Frank Hooey's treatment was a success story. However, it takes around 650 pituitary glands to produce the 2 to 3 grams of human growth hormone needed for the five year treatment and there simply aren't enough donors. Because of this, human growth hormone obtained from pituitaries costs far more than gold and only a small number of people benefited from the procedure.

However, the advent of genetic engineering has changed all this. Human growth hormone is a protein and once the gene responsible for its production was found and isolated, it was a relatively simple matter to insert it into laboratory bacteria and get them to synthesise the protein. The human growth hormone is then collected, checked for purity and given to people suffering from a shortage of it.

You might think that this is a perfect example of genetic engineering in action. Bacteria are being used to replace a missing human protein. As being 4 foot in height is manifestly disadvantageous, the procedure seems extremely useful and surely only someone with an extreme aversion to genetic engineering could object. However, the truth is more complicated.

Who uses genetically engineered human growth hormone?

There are three main categories of people who use genetically engineered human growth hormone: children who would otherwise suffer from pituitary dwarfism and end up only about 4 foot (1.2 metres) in height; children who would otherwise end up round about 5 foot (1.5 metres) in height; and sportspeople.

The first of these categories – children who would otherwise suffer from pituitary dwarfism and be only about 4 foot in height – are the ones for whom genetically engineered human growth hormone was originally intended. For such children, there is a lot to be said for genetically engineered growth hormone. Growth hormone obtained by the old method of extraction from the pituitary glands of countless corpses was almost unobtainable. Further, when it was available, it was occasionally contaminated by viruses. Over a dozen cases are known where children who had received human growth hormone extracted from pituitary glands went on to develop Creutzfeldt–Jakob disease – a rare and fatal condition caused by a virus that can infect human brain tissue.

However, there really aren't very many children with pituitary dwarfism, probably fewer than 1000 in most countries as only about 1 in 100 000 people are pituitary dwarfs. What there are, though, are millions of children who, though taller than pituitary dwarfs, are shorter than average. What has happened since the advent of genetic engineering is that some parents have started to take advantage of the available technology and have put their children onto programmes of human growth hormone injections, hoping that the children will end up taller. It might be supposed that this isn't a very serious problem. After all, many parents pay for their children to receive music lessons or tennis coaching. What does it matter if some parents pay for their children to receive human growth hormone treatment?

One problem is the cost of the procedure. A full course of treatment lasts between five and ten years and costs, at 1993 prices, up to $150 000.[8] A second problem is that no-one knows for certain if the treatment works. We know that pituitary dwarfs benefit, because the injected growth hormone replaces the missing growth hormone. But there are lots of reasons why children are below height and some paediatricians doubt that injections of human growth hormone will help in all cases. Actual findings are unclear. There are some data which suggest that injecting short children who are not growth hormone deficient does not increase their eventual height. There are other data, obtained by Genentech, the leading manufacturer of genetically engineered human growth hormone, which suggest that growth hormone injections do increase the height of both boys and girls.

A third problem is that once you start the treatment, you have to continue. As stated by the National Institutes of Health in the USA: 'Stopping growth hormone treatment in children who are not growth hormone deficient before they have reached their adult height may cause them to grow more slowly than they did before treatment. This may occur because taking the extra growth hormone causes the body to temporarily stop making its own growth hormone'.[9]

A fourth problem is that even if the full treatment does cause children to grow a few inches taller, the effects on a child's self-esteem and mental health are unknown. Maybe these will be bolstered. However, it has been suggested that quite the opposite may be the case. The injections may cause children to see themselves as abnormal, with subsequent loss of self-esteem.

A final problem with genetically engineered human growth hormone is that a number of independent studies have suggested a possible causal relationship between its long-term use and leukaemia. As a result, both Genentech and Eli Lilly, companies that produce genetically engineered human growth hormone, have changed their labelling to indicate this.[8]

It is worth bearing in mind that half of us are shorter than average. It is true that tall people benefit in all sorts of ways: research has shown that, other things being equal, tall people are more likely to be favoured at job interviews, while the taller candidate has won the great majority, 80%, of USA Presidential election campaigns. Surely, though, society should be challenging this bias in favour of the tall, rather than conniving with it by allowing parents to spend many tens of thousands of dollars in an effort to enable their children to grow a little taller. As Abby Lippman, a professor at McGill University and Chair of the Human Genetics Committee of the Council for Responsible Genetics, has suggested: 'Why not lower the hoops on a basketball court?'.[10] Diane Keaton, of the National Association of Short Adults, points out that being short has certain ecological advantages: 'We use less food and fiber. We take up less space'.[11]

A third category of people who are using genetically engineered human growth hormone are certain sportspeople. Some athletes have used the hormone in an attempt to increase strength, in much the same way as steroids are used illegally for the same purpose. In 1991, Dan Leggett, Compliance Officer at the FDA was quoted as saying: 'If the average jock could afford hGH, he'd pass up anabolic steroids'. The consequences of larger than normal doses of human growth hormone are still incompletely known, but it won't be surprising if it turns out that there are harmful medical consequences. In addition, the practice, if it works, is unfair on athletes who don't inject themselves.

Recently, genetically engineered human growth hormone has been given to a number of people with AIDS and to large numbers of people over the age of 50. There is considerable, though as yet largely anecdotal, evidence that human growth hormone can increase muscle strength, reduce fat deposition and reduce depression. No doubt time will tell whether we really have found the elixir of life or whether there are significant side effects.

Growth hormone from cattle: BST

Cattle produce growth hormone, just as humans do. Cattle growth hormone is called bovine somatotrophin or BST for short.[12] In a dairy cow, BST has two main functions, both to do with the nutrients derived from the cow's food. BST channels some of these nutrients into growth, and it channels some of them into milk production. The details need not concern us. Suffice it to say that BST is a natural hormone produced by cattle. During lactation, BST causes nutrients derived from the cow's food to be diverted to her mammary glands where they are used to make milk. It is this fact that has led to an extraordinary ten year battle over genetically engineered BST.

Why bother to make genetically engineered BST?

Because BST causes nutrients derived from a dairy cow's food to be diverted to her mammary glands, it was suggested, in the early 1980s, that injecting a cow with BST might increase her milk yield.

The chemical structure of BST was determined in 1973, and by 1982 genetically engineered BST had been made by incorporating the cattle gene for BST into the bacterium *E. coli*. A huge amount of basic and applied research has been carried out on genetically engineered BST by a number of companies including Elanco, Cyanamid and Upjohn, but especially by the multinational Monsanto.[13,14]

Injecting dairy cows with genetically engineered BST increases their milk yields by some 20%. Furthermore, it increases milk to feed ratios by some 15% – that is, the amount of milk made by the cow relative to the food she consumes goes up by around 15%. Monsanto argue that genetically engineered BST offers a number of significant advantages to dairy farmers. The main one is that by raising milk yields and increasing feed efficiency it raises profits. Further, the technology requires no capital investment: the farmer merely injects the cows every 14 days from the ninth week after calving until the end of lactation. This means that the system is, Monsanto maintains, 'immediately self-financing'.[15] In other words, there is no significant lag time between the introduction of the technology and the financial benefits reaped as a result of it. Finally, Monsanto points out that genetically engineered BST is virtually identical to the BST naturally produced by cows, differing by just one amino acid residue. BST is present in milk only in trace amounts and has no physiological effects on humans, as it differs in structure from human growth hormone. In any case, the minute amounts present in milk are digested and so do not pass into the human bloodstream.

Disadvantages of using genetically engineered BST

Despite the apparent financial benefits to farmers of using genetically engineered BST, a number of different arguments have been put forward as to why it should not be used. Four main ones can be proposed.

First of all, who wants or needs more milk? The number of dairy cows kept in Europe and the USA – the two regions where profitable sales of BST are most likely – has fallen over the last 20 years. This is because the demand for milk has failed to keep up with the dramatic increases that have already taken place in milk yields through selective breeding, feed concentrates and other agricultural practices. Indeed, Europe has seen a surplus of milk production for most of the last ten years, leading to the occurrence of so-called milk lakes. It is likely that the introduction of genetically engineered BST would lead to even more farmers being put out of business. The validity of this argument has been acknowledged, in part, by Monsanto. In some of their promotional literature they argue, rather confusingly, that: 'While it may seem paradoxical to be introducing a new technology into an industry where there is already surplus production, it should be stressed that BST is **not** intended solely to produce more milk. Instead it can be used either for this purpose, or to reduce the cost of milk production in a situation where extra milk is not desired – such as exists in a quota system'.[16]

Secondly, while it is true that genetically engineered BST itself almost certainly poses no health risks to humans, its use is linked to significantly raised levels of insulin growth factor-1 in the cow's milk. The consequences of this are still controversial. It has been argued that the presence of these high levels of insulin growth factor-1, which is chemically identical in cattle and in humans, may trigger premature growth in infants, breast development in children and breast cancer in women. While few scientists regard this possibility as a likely one, Ben Mepham, a physiologist working at the Centre for Applied Bioethics at Nottingham University, has argued that legalisation of commercial use of BST in the absence of more extensive information on these questions could lead to a deterioration in public health, not least if widespread rejection of milk were to result. He has also pointed out that public confidence in biotechnology would be helped by a more open system of regulation and by the use of 'blind trials' in experimental work, as is the norm in the pharmaceutical industry.[17]

Thirdly, does the use of BST injections harm the cow's health? Even the manufacturers of genetically engineered BST accept that its use may increase the incidence of mastitis, cystic ovaries, disorders of the uterus, retained placentas and other health problems, including indigestion, bloat, diarrhoea and lesions of the knees.[14] In addition, its use may result in permanent swellings up to 10 cm in diameter at the injection site.[14] Mastitis, as many women know all too well, is a painful inflammation of the mammary glands. It has the same effects in cows as in humans. (As an ironic twist to the story, research has been undertaken in the hope of producing genetically engineered cattle resistant to mastitis.) Mastitis is commonly treated by giving infected cows antibiotics. Some concern has been raised at the consequences of this for human health, though these fears may be exaggerated as antibiotics have been used on farm animals for decades. A related point is that BST injections possibly put even more pressure on a cow's health in countries where farmers have little or no access to high concentrate feeds.

Fourthly, while everybody knows that dairy farming is big business, for many people the thought that cows will be artificially stimulated by biweekly injections of genetically engineered BST for most of their lives is somehow off-putting. True, genetically engineered BST is almost identical in structure to natural BST, but to some people it seems wrong that it should be used to boost a cow's BST levels beyond what is normal. Is its use analogous to the force-feeding of geese to produce pâté de foie gras? For some people, milk still retains a special aura of freshness and naturalness, perhaps because we all start our lives, once born, by living off milk. This image is tarred by the use of genetically engineered BST. It may be hard to reconcile a belief, albeit a naive one, that milk is a 'natural' product with the recognition that genetic engineering is being used to direct the process (see Figure 5.1).

Igor Kopelnitsky,

Figure 5.1.

The current legal position

By 1994, genetically engineered BST had been licensed for use in a number of countries including South Africa, India, Mexico, Brazil, the former USSR and, in February 1994 – and after a long battle – the USA. Endless debates have taken place within the European Union but in December 1994 agriculture ministers from European Union countries agreed to continue the ban on its use until the year 2000. Britain alone advocated lifting the ban, but was outvoted 11 to 1. A UK government consultation in 1994 showed that three-quarters of the organisations that responded, including consumer groups, animal welfare organisations, environmental pressure groups, organisations representing farmers and four of Britain's largest supermarket chains, wanted the ban to remain in place.[18] The organisations that wanted the ban lifted were mainly biotechnology companies and scientific institutions. One of the most remarkable features of the lifting, in the USA, of the ban on the use of genetically engineered BST was that the Food and Drug Administration produced guidelines stating that any company proclaiming that its milk was produced without the use of genetically engineered BST would have to carry a long statement explaining that there is no advantage to BST-free milk.[19] The first two American dairies that advertised their milk as 'hormone-free' were promptly sued by Monsanto.[20]

Genetically engineered viruses to control insect pests

The three examples we have considered so far – genetically engineered human insulin, human growth hormone and BST – have been examples of substances made by genetically engineered bacteria and injected into individuals – humans or cows. A very different example of the use of genetic engineering is provided by the case of genetically engineered viruses, used to control insect pests.

The control of insect pests

Worldwide, somewhere around a third to a half of all agricultural production is lost to pests and diseases. Insect pests are amongst the most important pest organisms and chemical insecticides are widely used to reduce their effect. Since the early 1980s, there has been a resurgence in the use of so-called 'biological control'. Biological control is where one organism is used to control another. Biological control was practised by the Chinese some 4000 years ago when ants were used to kill leaf-eating insects, thus protecting valuable crops such as oranges.

Perhaps the most famous example of biological control was the use of a moth called *Cactoblastis cactorum* to control the prickly pear cactus (*Opuntia*) in Australia. The prickly pear cactus is not native to Australia. It was introduced there by a certain Dr Carlyle who grew the plant in his garden because he thought it looked nice and might do well. It did. By 1910, the cactus was spreading at a rate of over 300000 hectares a year. The Australian government responded by trying to find an animal that would eat it. In 1925, eggs of the moth *Cactoblastis cactorum* were introduced into Australia from Argentina. These proved a great success. Within a few years almost all the prickly pear cacti had been destroyed.

Why use genetic engineering?

Genetic engineering is being used to try and improve the effectiveness of biological control. Some of the most thorough work to date has been carried out by David Bishop and Jennie Cory in the UK.[21] Bishop and Cory work on viruses known as baculoviruses. Baculoviruses attack insects. Natural (non-genetically engineered) baculoviruses have been used for decades to protect crops simply by spraying the viruses onto crops infected with insect pests such as aphids or caterpillars. One of the great advantages of baculoviruses is that they have absolutely no effect on vertebrates. This means that there is no danger to humans from eating crops treated with them. Given the widespread concern over the effects of chemical pesticides both on human health and on animals such as sparrow-hawks and fish, baculoviruses hold out the hope of safe, effective insect control.

However, one problem is that natural baculoviruses are relatively slow in their effect. This is a common problem with biological control agents. They need time to multiply up in the field. Because of this, Bishop and Cory are carrying out a series of experiments in which insect-specific toxins or insect-specific hormones are introduced into the viruses through genetic engineering. The idea is that the virus will be even more effective at killing the insect pests. Either it contains a toxin that poisons the insect or it contains an insect hormone that causes the insects to fail to develop normally.

So far so good. However, in 1994 a major row broke out over the research.[22] The research group had been given permission to spray some of the genetically engineered virus in a field near Wytham in Oxfordshire. However, this field is only 100 metres from a nature reserve inhabited by rare moths. The viruses had been modified by having a scorpion gene inserted into them – the gene making a venom lethal to insects. Although Bishop and Cory had already, in 1993, carried out a very similar experiment, the new experiments sparked off a major controversy amongst scientists and led to articles with such headings as 'Will the scorpion gene run wild?'.[22]

On the one hand, the UK government's Advisory Committee on Releases to the Environment, which evaluates all proposals for outdoor tests with genetically engineered organisms, approved the experiment. On the other hand, a number of ecologists have subsequently expressed concern. Might some of the viruses escape from the test site? David Bishop counters this fear by pointing out that the enclosures that surround the experimental plots consist of tough, fine netting designed both to stop the wind blowing the virus out and to prevent animals from getting in. This netting has been extensively used and proved effective.

The major fear is the effect that the viruses might have if they do come into contact with native insects. In the 1993 experiments, Bishop and Cory found that the genetically engineered viruses killed caterpillars of the cabbage looper moth faster than the normal virus. As a result, the genetically engineered virus afforded significantly greater protection to the cabbage crop on which the cabbage looper moth caterpillars were feeding. What is not really known is the effect that the genetically engineered virus might have on native moths. There is genuine academic disagreement about which moth species are attacked by the natural, let alone the genetically engineered, virus. Work carried out by Bishop suggests that only a very small proportion of British moths are susceptible to the virus. However, Alan Wood, who studies baculoviruses at the Boyce Thompson Institute for Plant Research in New York, believes that these figures are too low and points out that there have even been reports of the virus infecting termites and beetles.[23]

A related fear is that the use of the genetically engineered virus might allow the scorpion gene to be picked up by other viruses. Again, academic disagreement exists as to how likely this is in the field, though it has been demonstrated in the laboratory.

Other issues

The row over Bishop and Cory's research with genetically engineered viruses has focused exclusively on the issue of safety. As we saw in Chapter 2, it is virtually impossible to predict the risks of an action before the action is carried out. Bishop and Cory's research has to be undertaken if we are to evaluate the risks of the procedure – though it would perhaps be wiser to carry it out rather further from a nature reserve.

However, the use of such viruses to control crop pests raises issues that go beyond safety. Suppose we knew that the widespread, commercial use of genetically engineered baculoviruses would increase the value of crops by, say, £10 million a year but carry a 1% chance each year of leading to the extinction of a native species of moth. How do we decide whether, given this clear-cut scenario – and in life the alternatives will never be as definite as this – we should legislate for the use of genetically engineered baculoviruses or ban them?

Frost damage caused by ice-nucleation bacteria

As everybody knows, water freezes at 0°C. Except that, rather surprisingly, it often doesn't. Water can exist in a supercooled state. Supercooled water is still liquid but has a temperature below 0°C. Supercooled water may freeze if so-called ice nuclei are present. Ice nuclei are tiny particles that allow the formation of ice crystals to get started. Once ice crystals start to grow in supercooled water, they quickly spread, causing the water to freeze solid. However, in the absence of these ice nuclei, the water may remain liquid instead of freezing, even at temperatures well below zero.

This property turns out to be of great commercial significance. Many crops are damaged by the formation of ice crystals within the plants. Particularly susceptible are the flowers of fruit trees and the fruits themselves. A single late frost, as gardeners and farmers both know, can ruin an entire year's crop. In nature, the formation of ice crystals within plants is often triggered by the growth of bacteria on the outside of the plants. Some of these bacteria have proteins on their surfaces that, unfortunately from the farmer's point of view, are particularly effective triggers of ice nucleation. For this reason, these bacteria are known as ice-nucleation bacteria. In the absence of ice-nucleation bacteria, plants may have an internal temperature as low as $-5°C$ without freezing.[24]

Preventing ice nucleation

In the mid-1980s, the damage caused by frost injury to crops in the USA was estimated at $1 billion.[25] Frost damage is big business. Farmers have tried various ways of reducing its impact. One approach is to warm the air or insulate the crops: straw may be burned on cold nights and fans used to keep warm air circulating. A different tactic is to try to get rid of the bacteria that encourage the formation of ice crystals by spraying the crops with bactericides that will kill the bacteria. Another approach is to spray the plants with bacteria that inhibit the growth of ice-nucleation bacteria.

It is this final approach – spraying the plants with bacteria that inhibit the growth of ice-nucleation bacteria – that has led to the use of genetic engineering. The history of this controversy has been surveyed by Sheldon Krimsky, a professor at Tufts University, USA.[26] Krimsky is also on the Board of Directors and the Editorial Board of *GeneWATCH*, a Bulletin critical of genetic engineering, produced by the Council for Responsible Genetics.[27]

The bacterium most responsible for ice nucleation is called *Pseudomonas syringae*, or *P. syringae* for short. One of the proteins made by *P. syringae* sits on the outside of the bacterium and causes ice crystals to form as soon as the temperature drops below about −1°C. (As an aside, there is a significant market for this protein, which can be used to help make snow for ski slopes![28]). For the bacterium, the protein is thought to be useful because damaged plant cells leak nutrients; the bacterium can probably absorb and use these nutrients.

Research by Steven Lindow at Berkeley, California in the early 1980s showed that if *P. syringae* is genetically engineered so that the gene responsible for the ice-nucleating protein is removed, it loses its ability to cause ice nucleation at relatively high temperatures. Such forms of *P. syringae* are known as ice-minus strains. Formal submissions were then made by a company called Advanced Genetic Sciences to the various USA regulatory authorities. The submissions called for field trials to take place so as to see whether genetically engineered ice-minus strains of *P. syringae* enable crops to suffer less frost damage.

So far the story may sound like a relatively uncontroversial and useful example of genetic engineering. Note that nobody is putting a foreign gene into *P. syringae*. All that is happening is that one of its own genes is being removed. As has been pointed out: 'This example involves only the modest kind of genetic manipulation. The bacterial genome has nothing added to it and has just been subjected to a deletion of a kind that could, and doubtless does, occur spontaneously at a low frequency in nature'.[29]

However, the technique will only work if the genetically engineered bacterium succeeds in establishing itself in such large numbers on the crop that it effectively replaces the natural form. This is the same situation as the genetically engineered baculoviruses we looked at earlier, which may help reduce pest damage (p. 111). However, every effort was made in Bishop and Cory's research with baculoviruses to prevent escape of the genetically engineered microorganism from the experimental plots through the erection of appropriately designed netting. No such efforts were made by Advanced Genetic Sciences.

Objections to ice-minus bacteria

In June 1983, the USA National Institutes of Health, through its Recombinant DNA Advisory Committee, formally approved field trials of ice-minus strains of *P. syringae.* A number of individuals and organisations protested at this decision and Jeremy Rifkin instigated lengthy legal proceedings to stop these and other trials of genetically engineered organisms. One of the main arguments raised by Rifkin and his Foundation on Economic Trends against the National Institutes of Health decision was that it had failed adequately to address the possible impact of ice-minus bacteria on the climate. It is known that rainfall is triggered by nucleation events in the atmosphere. As Steven Lindow himself has written: 'Yet the importance of ice nucleation-active bacteria in contributing ice-nuclei active at warm temperatures in the upper atmosphere for the formation of rain and snow is as yet largely unexplored. The role of these bacteria may be potentially of critical importance in climatology studies'.[30]

Rifkin pointed out that the National Institutes of Health Recombinant DNA Advisory Committee – the body that approved the field trials – contained no ecologists, botanists or population geneticists. Yet it is precisely in these areas that expertise is needed if the risks of ice-minus *P. syringae* are to be evaluated. Rifkin's 1983 lawsuit was successful and a legal battle then ensued, and lasted several years, between those who wanted the trials to go ahead and those who didn't. Advanced Genetic Sciences' case wasn't helped by the disclosure that, in 1986, it carried out a small-scale unauthorised release of a strain of ice-minus bacterium. Nevertheless, the trials were eventually approved and the first ones took place in California on 24 April 1987. To the best of our knowledge, California's climate has not changed as a result.

How safe are ice-minus bacteria?

Science cannot always answer the question 'How safe is something?'. In the case of ice-minus bacteria, the scientific community itself polarised to a certain extent. On the one hand, there were molecular biologists and scientists familiar with the deliberate introduction of strains of nitrogen-fixing bacteria into the environment – a practice which has safely taken place for decades.[31] These scientists generally saw little cause for concern. On the other hand, there were ecologists aware of how little is understood of the fundamental principles of ecology. On balance, few scientists, whatever their particular discipline, really believed that even the widespread commercial use of ice-minus *P. syringae* would cause any significant alteration in the climate. But then scientists have often been proved wrong, and ecology and climatology are not yet as well understood as some other branches of learning.

We can't yet be certain, or even *very* confident, that the use of ice-minus *P. syringae* is safe. Most of what had been learned by 1995 about the risks of introducing genetically engineered microorganisms into the environment suggests that the risks are very low.[28,32] What many people find disconcerting, though, is that both in the USA and in Europe regulations over the testing and use of genetically engineered microorganisms in the field are being relaxed.[33] Changes in the USA regulations meant that a 1994 field experiment involving the release of a baculovirus that had had one of its genes deleted proceeded without the study even having been referred to the USA Environmental Protection Agency. Bill Schneider, the Biotechnology Co-ordinator at the Environmental Protection Agency's Office of Prevention, Pesticides and Toxic Substances, was quoted as saying: 'We are not loosening things up, but we are allowing people to test some low-risk things without telling us'.[34]

In the light of this relaxation of legislation, it is disconcerting to discover that a recent study in the Netherlands of companies mailing microorganisms through the post revealed that not one laboratory packaged its samples correctly.[35] The study involved scientists at the Dutch National Institute for Public Health and Environmental Protection requesting samples of genetically engineered micro-organisms from a total of ten laboratories in Holland, the USA, Australia and Singapore. None conformed to the UN or Dutch specifications. The Dutch national bacteriological laboratory get 1000 samples in the post each year, most being bacteria sent from hospitals for identification. Broken, leaking parcels were at one point so frequent that the laboratory has sent out free, appropriate mailing containers to hospitals. Even so, it still gets about five broken containers in the post each year.

Vegetarian rennet

The final example we shall consider of the use of genetically engineered microorganisms raises a different set of ethical issues from those we have addressed so far. It concerns the making of cheese.

Cheese has been made by people for at least 5000 years. The fundamental principles have changed little over the millennia. During cheese making, a number of substances are added to sour milk. One of them is rennet. Rennet is a crude extract of enzymes, of which much the most important is chymosin, also known as rennin. These enzymes act on a milk protein called casein. Their effect is to cause the milk to form a soft curd, also known as junket. Without rennet, most cheeses cannot be made.

Traditionally, rennet has been obtained from the stomachs of young calves (or piglets, kids, lambs or water buffalo calves). Rennets can be of vegetable origin, but, until recently, by far the most important source was young calves. Calves' stomachs were ground up in salt water – ten of them being required for one gallon of rennet.[36] Of course, calves' stomachs contain a lot of stuff in addition to rennet, so the purity wasn't very high. However, the gene for calf chymosin has now been inserted, by genetic engineering, into a yeast, which produces a ready supply of chymosin in commercial quantities. As a result, the use of rennet obtained directly from animals has greatly decreased. In addition, genetically engineered chymosin is cheaper than traditional rennet and considerably purer.

The original gene used in the genetic engineering of the yeast came from an animal source. However, the Vegetarian Society of the United Kingdom decided to endorse genetically engineered chymosin on the grounds that its use would significantly decrease the slaughtering of calves. Accordingly, manufacturers that produce cheese through the use of genetically engineered chymosin can put a V-symbol on the cheese and state that the product is suitable for vegetarians on the packaging. It is also approved by Muslims and Jews.

In the UK, the Co-op supermarket chain was the first retailer to announce a policy on the genetic engineering of food. It has produced a free leaflet called *The Right to Know: Your Guide to*

Genetic Engineering, which is available from its stores and also from its Customer Relations Department.[37] Key points within the policy are:

- No food product containing modified human genetic material will be sold by the Co-op
- Co-op Brand products will not contain vegetables or fruits which have been modified with genetic material from animal sources
- All Co-op Brand products known to contain modified genetic material from non-related species will be clearly labelled

By 1994 the only product in the Co-op's range that had used genetic engineering was Co-op vegetarian cheese. This product is labelled 'Produced using gene technology, and so free from animal rennet'.

One important factor in the Vegetarian Society's decision to approve genetically engineered chymosin was the fact that, in practice, the rennet actually used by cheese manufacturers will not contain the original calf chymosin gene, but copies of it. However, something of a diversity of views exists on this point among vegetarians:

> The concept of copying rather than direct transfer was important in our decision to allow chymosin manufacturers to use the V-symbol. However, vegetarians are obviously not a homogeneous group and will have divergent views on this issue. Some will accept the copying stage as meaning a host containing a gene from an animal source is acceptable whilst others will view its animal origin as meaning that the product is not acceptable.[38]

By 1994, approximately one half of the market for rennet worldwide was being supplied by genetically engineered chymosin. It is tempting to see this example of genetic engineering as a way of saving the lives of the millions of four- to ten-day-old calves that, until recently, were killed for their rennet each year. The reality, though, is that these calves continue to be produced so as to keep their mothers, dairy cows, producing milk. Female calves generally themselves become dairy calves. Male calves are usually reared for meat, either as veal calves or as beef cattle.

WHAT'S WRONG WITH GENETICALLY ENGINEERING MICROORGANISMS?

The genetic engineering of microorganisms does not at first sight look to be a subject capable of arousing much moral concern or controversy. After all, we cannot even see these organisms (without a microscope) and so do not normally devote much time and effort to considering their welfare. However, the previous six case studies have already shown that the issues raised are far from being purely scientific and, in fact, often contain a clear moral dimension. Following the distinctions drawn between 'moral' and 'ethical' in Chapter 3 (pp. 45–7), we can now examine the possible areas of moral concern highlighted by these case studies and subject them to some ethical scrutiny. Where relevant, religious and theological considerations will also be included.

The range of moral concerns implicit in these case studies is extremely wide, too wide in fact to be dealt with in one section of a single chapter. Therefore, as each chapter in Part 2 will conclude with a separate ethical analysis, certain topics will be reserved for certain chapters, where they can be given fuller treatment, rather than attempting to deal with every ethical aspect of every case study in every chapter, which would inevitably lead to repetition. In this chapter, for example, the use of human growth hormone poses serious ethical and theological questions about the extent to which human differences should be regarded as 'defects' to be remedied or even prevented by genetic engineering, while BST has aroused much moral unease on grounds of animal welfare. Although these examples involve the genetic manipulation of microorganisms, the ethical and theological issues will be more conveniently dealt with not here but in Chapters 7 and 8, where the focus will be upon animal and human applications of genetic technology.

This still leaves plenty of ethical work to be done in this chapter. The genetic engineering of microorganisms in general and the case studies in particular illustrate graphically many of the basic distinctions and arguments introduced in Chapters 3 and 4. Following the framework developed there, we can see that moral concerns arising from the case studies may fall into either the intrinsic or the extrinsic category.

Respect for nature

Intrinsic concerns that the technology is wrong in itself and so ought not to be used can of course be expressed about *any* application of genetic engineering, irrespective of the possible advantages and disadvantages in particular cases. We offered a general ethical and theological assessment of such arguments in Chapters 3 and 4 and pointed out some of the weaknesses and confusions that they may suffer from. These points will not be repeated in detail in this or the following three chapters. However, intrinsic objections to the genetic engineering of micro-organisms carry with them some more specific implications which help to clarify and further develop the general comments already made.

Moral or religious qualms about 'interfering with Nature' or 'playing God' are strictly speaking as relevant to the manipulation of laboratory bacteria to produce insulin as they are to work on animal or human genetic material, for DNA ('the stuff of life') is being genetically altered in all such cases. Despite this logical equivalence, however, intrinsic objections do seem, intuitively at least, to carry less weight when the object of the manipulation is an invisible microbe than when it is a pig or a person.

If we probe this intuition further, we can see that important questions have to be asked about intrinsic objections to genetic engineering that are based on an alleged lack of respect for 'Nature' or for 'life' itself. To put the problem at its simplest, what are the boundaries of 'Nature' or of 'life'? Do they extend to the world of microorganisms? No doubt the microbiologist would maintain that they do, but this scientific categorisation creates serious difficulties for those who wish to attribute moral or religious value to *all* of 'Nature' and to *all* 'living' beings and who consequently object to any genetic interference with them.

Do invisible organisms that appear to possess far less 'consciousness' even than insects automatically merit our 'respect', even when they may at times be responsible for our suffering and death? What would in fact constitute 'respect' for a microbe? The theologian and missionary Albert Schweitzer is well known for his passionate advocacy of 'Respect for all Life', but we would be surprised to learn that he felt or showed much respect towards those living microorganisms which brought disease and death to some of those for whom he was caring. Should doctors and medical researchers respect harmful viruses and bacteria (in any sense other than appreciating their dangerous potential and taking appropriate precautions)?

Posing such questions to those who might feel that genetically engineering even microorganisms is somehow intrinsically wrong is in effect another way of underlining the confusion that can surround the terms 'Nature' and 'natural', as was demonstrated in Chapters 3 and 4. Moral decisions cannot be made simply by pointing to or reading off what exists in the natural world. We make moral decisions *about* that natural world and how we ought to behave towards it: for example, that we ought not to torture cats or destroy rainforests, and that we ought to build flood barriers or seek to eradicate the AIDS virus. The mere existence of organisms or phenomena in the natural world provides us with no automatic moral directives about how we should act.

Similar difficulties face religious objections of an intrinsic kind to the genetic engineering of microorganisms, based on the belief that all elements of the natural world, created by God, are by definition 'good' and not to be interfered with. We showed in Chapter 4 that this is an extreme position and that alternative views of the moral and religious status of creation and the natural world are widely held. If the claim of intrinsic wrongness is made on religious grounds, however, the implications and problems are similar to those arising from the non-religious version that we have just discussed.

What sense can be made of the view that microorganisms possess intrinsic religious value, when the sole function of some of them is to cause harm to other forms of life, including plants, animals and human beings? Just as moral decisions about what to do cannot be deduced by simply observing the natural world, so is it impossible by such methods to reach any religious conclusions about God's goodness and will. The apparent 'cruelty' of Nature and the problem of 'how could a benevolent God create such a cruel world of carnage and bloodshed'[39] puzzled many nineteenth century theologians. They appreciated the difficulty of finding divine goodness exemplified in the cat playing with its mouse or in the ichneumon fly injecting its eggs into caterpillars, thus initiating a slow death of parasitic ingestion, which has been likened to the ancient torture of drawing and quartering.[40] Such theologians today would doubtless be equally troubled by the idea of divine goodness being embodied in some crippling and deadly microorganisms.

In addition to these problems confronting anyone who tries to condemn the genetic engineering of microorganisms on intrinsic grounds, there is the further difficulty of specifying what is distinctively 'unnatural' about this technology as opposed to others. To use some of the examples provided by the case studies, why should the genetic engineering of human insulin or growth hormone be seen as any more 'unnatural' than many other methods of producing drugs and pharmaceuticals? Deriving insulin from the pancreata of cattle and pigs can hardly be called a 'natural' process. Obtaining rennet for cheese-making from the stomachs of young calves seems no less an interference with Nature than genetically engineering a yeast to produce chymosin. The highly mechanised techniques of modern dairying can be accused of 'unnaturally' treating cows as mere 'milk machines',[41] whether or not BST is used to boost supplies.

As we pointed out in Chapter 3 (p. 60), practically every element of our modern Western life-style could be labelled 'unnatural' or an interference with Nature, including all of medicine and agriculture. Objections to the genetic engineering of microorganisms on these grounds, if they are to be consistent, should also be extended to all medical and agricultural practices and to a vast range of other human activities besides, necessitating, in effect, a complete rejection of and withdrawal from the modern world. That is, of course, a possible moral position to adopt (in theory at least), but those who feel concern about genetic engineering on intrinsic grounds should be aware of the logical implications of their viewpoint.

Safety, costs and benefits

In practice, there are probably not many people who would object
to the genetic engineering of microorganisms on the grounds that it
is intrinsically wrong. Far more widespread are concerns about
safety, and the six case studies have drawn attention to a variety of
risk factors. All of those case studies in fact, except the last, show
that concerns about possible risks have been expressed, most
notably in the cases of ice-minus bacteria and of pest control.
These two applications of genetic technology provide good examples of possible 'Doomsday scenarios', as envisaged by critics such
as Rifkin, who was quoted in Chapter 3 warning against playing
'ecological roulette': '. . . while there is only a small chance of . . .
triggering an environmental explosion, if it does the consequences
can be thunderous and irreversible'.[42]

However, we also saw in Chapter 3 that the theoretical possibility
of an activity leading to catastrophic consequences does not
necessarily mean that that activity is ethically unjustifiable, for such a
principle would in effect outlaw most if not all scientific research
(p. 55). In addition, it was pointed out that choosing the 'safer' option
in such cases is not as easy as it sounds, because abandoning an
allegedly 'risky' line of research today may have catastrophic
consequences tomorrow. It is not inconceivable, for example, that
research aimed at enabling crops to withstand lower temperatures
might lead to developments that might help to avert future disaster in
the event of significant climate changes. Similarly, biological pest
control might play a part in preventing future famine and starvation.

Ethical judgments about the acceptability of risk, then, must
attempt to weigh the likely costs and benefits on a case-by-case
basis, but the case studies show how difficult this is where controversy exists over the magnitude and extent of the risk, the likely
short-term and long-term consequences and the methods of
assessment. One recent writer on this subject has claimed: 'Utilitarian calculation of this kind has one stunning disadvantage . . . it
is wholly incalculable'.[43]

Even if we do not wish to go quite as far as that, the problems are
clearly considerable. This is probably less so in a case such as
human insulin, where the risks and benefits are limited to a single

dimension and are, at least in principle, calculable than in a case such as BST, where a wide variety of possible costs has to enter into the ethical arithmetic.

BST is in fact a perfect example of an application of genetic engineering that raises a complex mixture of moral concerns and so deserves more detailed investigation at this point. The case study has already shown that these concerns include queries about the possible risks involved both for the consumer of milk and for the cow that produces it, about various aspects of animal welfare, about the socio-economic threats to small farmers and about the 'unnaturalness' of the process and the resulting product. Given this wide array of emotive issues, it is hardly surprising that the introduction of BST has been far from smooth and has encountered heated opposition. In the USA hard-hitting television commercials have been produced by environmental pressure groups demanding 'What are they doing to our milk?'. In view of the overproduction of milk in Europe and North America, many critics of BST have summed up their case with the question, 'Who *needs* more milk?'.

We should, however, always be wary of arguments and claims about 'need', for they are never as simple or straightforward as they may at first appear. The statement 'we need X' cannot be a plain statement of fact in the way that 'we want X' or 'we lack X' is. To say 'we need X' is both to *describe* our situation in lacking X (or being likely to lack it) and to *prescribe* that we ought to have X in order to achieve some further end that is thought desirable. So any questions about needs inevitably involve values and value judgments and cannot be answered conclusively just by pointing to sets of facts. Therefore, the question, 'Do we need BST milk?', is not a factual question but rather another way of presenting the problem of weighing one set of values or priorities against another. So if the resulting increase in productivity and in understanding of new techniques is judged to be more desirable than the possibly damaging social effects of those increases, then the answer will be 'Yes, we do need it'. If the balance of desirability is judged to be otherwise, the answer will be 'No, we don't need it'. Formulating the problem in terms of 'need', then, cannot by-pass the ethical questions about BST.[44]

These questions cannot be answered simply by looking at the financial balance sheet and arguing (along with the Ontario Milk Marketing Board): 'If you can produce the same amount of milk from fewer cows with less labour, more net profit, and get more time off, why shouldn't you?'.[45]

Trying to compare even the purely economic costs and benefits of BST raises the issue of fairness, which has long been a problem for utilitarianism (p. 58). It is possible that a large number of consumers might save a small amount of money through buying cheaper BST milk, while a small number of farmers might be displaced and lose their whole livelihood. What would count as a fair distribution of these costs and benefits?[46] Clearly, small farmers could be significant losers from BST as could also be the cows, who might (it is claimed) suffer in a variety of ways. The complex ethical arithmetic here is summed up by the American moral philosopher, Gary Comstock, as follows:

> I have argued that the costs and benefits of bGH [the alternative name for BST] will not be distributed fairly, because the potential costs will be concentrated on a group of innocent humans and animals, while the benefits will be thinly spread and fleetingly enjoyed. The costs are likely to include catastrophic harm for farm families and rural communities. To inflict such harms on innocent people is wrong, even if it is only a small number of people who are harmed and even if it will lead to large aggregated benefits. I have also argued the low-level, long-term, loss of freedom and reasonably stimulating surroundings will harm cows, as will the use of bGH by factory farm managers who do not pay close attention to the differences among their animals.
>
> In my moral universe, the interests of cows count. So do the rights of innocent individual humans not to suffer catastrophic harm. . . . I fear that the costs and benefits of bGH probably will not be distributed fairly.[47]

Clearly there are a number of serious ethical question marks about BST, though the ethical weighting to be placed upon the various costs and benefits is inevitably controversial, as is the general conclusion that Comstock draws regarding BST:

In short, bGH threatens to undermine the common good not simply of the dairy farmers it will displace, but of us all. It promises, in a small way, to undermine our general well-being.[48]

However, even if we might wish to make the ethical judgment that Western agriculture does not 'need' this sort of technology today, that is not to say that other agricultural systems might not 'need' it (or something like it) either now or at some future date. The chairman of a large Indian agricultural co-operative, for example, has claimed, 'To constrain production in a world of hunger is wrong, even immoral'.[49]

As with many applications of genetic engineering, a distinction may be drawn on ethical grounds between the possible undesirability of using the technology now in a particular economic and social context and the possible desirability of developing that kind of technology for use in other contexts now or in the future, where the ethical arithmetic could work out differently.

Labelling

BST and chymosin rennet highlight a further issue of ethical significance that will also relate to some of the case studies to be presented in Chapters 6 and 7 but which can for convenience be introduced here.

When trials of BST were extended from research institutions to commercial farms in the UK in 1987–8, the sale of milk from these trials was authorised into the pool of normal milk distribution. The BST milk was not kept separate from other supplies and the identity of the trial farms was not revealed. Consequently consumers had no way of knowing whether they were drinking BST milk or not. In the case of chymosin rennet, while producers and retailers are under no legal obligation to declare that genetic engineering has featured in the process, one major supermarket chain in the UK has publicised its belief that this information ought to be provided to shoppers and has acted accordingly.

These examples raise the general question of labelling food and drink products that involve genetic engineering and of the consumer's 'right to know'. The ethical issue here can be stated quite simply. If it is possible (as it clearly is) to have moral or religious objections to genetic engineering on intrinsic or extrinsic grounds, should not individuals have the right to choose not to buy products that involve these techniques in some way? And how can they have that right unless such products are clearly labelled? On the face of it, this argument seems unassailable – at least if we favour respecting people as autonomous moral agents, entitled to make their own moral decisions.[50]

However, objections have been raised about the practicality of labelling products in this way. For example, how could the complex scientific techniques used be made comprehensible to the majority of consumers? What is the point of such labelling if the product is identical with one that has not involved genetic engineering? If minute amounts of say BST milk or cheese made from chymosin rennet are used in other food products (e.g. a pizza), should these products also be labelled? If the range of foods involving some forms of genetic engineering increases as predicted (to include many basic cereal crops, for example), will not specific labelling become so widespread as to be unnecessary or even meaningless?

We seem to have here a clear conflict between ethical and pragmatic demands – though of course nothing that is in practice *impossible* can qualify as an ethical requirement, because claiming that something *ought* to be done implies that it *can* be done. One possible compromise might be to adopt what the Polkinghorne Report on the Ethics of Genetic Modification and Food Use refers to as the 'de minimis' principle (i.e. there needs to be no undue concern about very small matters): 'This is not inconsistent with the approach found in major ethical systems. All such systems normally incorporate a principle of "best endeavours" in the knowledge that absolutes are impossible to observe . . . It seems to us that it will be quite unrealistic to label every last element of modified food in every product in which it may be incorporated'.[51]

However, the Report considered that only genes of human or animal origin were 'ethically sensitive' enough to require labelling in foods and that this should apply only to foods *containing* such genetically engineered material, as opposed to being *products* of genetic engineering. None of the examples discussed in this chapter would, therefore, fall into this category.

These and other issues raised by the Polkinghorne Report will be examined further in Chapters 6 and 7 and in Part 3. On the specific question of labelling products such as BST milk and chymosin cheese, however, there does seem to be, for the reasons outlined earlier, a strong ethical case at the present time for giving consumers the right to refuse products about which they might feel serious moral or religious concern, however unfounded others might judge that concern to be.

Conclusions

Some of the main ethical questions raised by the case studies have now been explored. We have tried to tease out the main issues and principles at stake and to give some indication of how they might be approached. The ethical arithmetic appears clearer in some cases than others (though in none is it uncontroversially obvious). Our main aim in this and the following three chapters is to encourage the reader to draw his or her own informed ethical conclusions.

Even in this apparently less emotive area of engineering microorganisms, we have seen that there exist wide disagreements and no easy answers – a situation that makes it inappropriate to conclude this chapter with a set of our own dogmatic conclusions. In each chapter in this section, we shall, however, try to indicate where we stand at present on some of the issues, while emphasising that often the only useful and rational conclusions on such subjects are provisional ones.

It seems difficult to maintain that the genetic engineering of microorganisms is intrinsically wrong and we have criticised that view on a number of grounds in this chapter, following on from the more general criticisms developed in Chapter 3. Extrinsic concerns about safety need to be given more weight, particularly in view of the wide range of possible risks that have been mentioned in the case studies. Where expert opinion is divided about the level of risk, a cautious step-by-step approach is clearly called for. The potential risks and benefits in each case need to be weighed in order that responsible judgments may be made about what levels of risk are ethically acceptable. More will be said about this process in the following chapter. Our provisional judgment about the case studies presented here is that only in the case of BST does the ethical arithmetic suggest that the various costs and risks outweigh the benefits at the present time. However, genetically engineered human growth hormone can be misused, while it is too early to be sure whether the use of genetically engineered viruses to control insect pests is acceptable.

6

The genetic engineering of plants

Of course it will work. Give a scientist enough time and money and he can do anything.
Ken Barton, Vice-President of Research and Development at Agracetus[1]

It is not possible to quantify the direct ecological risks of the release of genetically engineered plants and we have very little knowledge on which to base our choice. The danger of genetic pollution is a very fundamental assault on nature. The dangers are irreversible, have the potential to multiply and would leave an unacceptable legacy to future generations.
Sue Mayer, Greenpeace[2]

Introduction

Plants form the basis of our diet. Most people in the world, either through choice or the lack of it, eat little or no meat. Even those with a diet high in animal products rely on plants, as most of the animals people eat, for example sheep, cattle and pigs, live mainly off plants. Even carnivorous fish such as mackerel ultimately rely on the countless, tiny plants that make up the phytoplankton that floats in the oceans. So plants matter because we need them for food. We also give them to one another as presents, grow them in our gardens and admire their beauty in the wild. Plants provide the oxygen in the atmosphere and significantly affect the climate. Without plants there would be few animals and certainly no humans.

As we reviewed in Chapter 1, farmers have bred plants for some 10 000 years. Today's wheat, even before the advent of genetic engineering, differs from its ancestors in so many ways that it is classified as a different species. Now genetic engineering offers the possibility of even greater change.

The genetic engineering of plants has lagged behind the genetic engineering of microorganisms by several years. One problem has been transferring foreign DNA into plant cells. This only became routine in the 1990s, a full decade after genetically engineered human insulin went on sale.[3] A second problem is that plants have much longer generation times than microorganisms. It takes time to get enough genetically engineered plants to set up field trials, and these field trials themselves typically take several years to carry out and evaluate fully. For these reasons, there are as yet few examples of genetically engineered plant products on sale.

In years to come, there is every likelihood that genetic engineering will revolutionise plant breeding. Often the genetic engineering of crops is held up as a potential world-saver. Perhaps it will enable us to feed the extra 250 000 people alive each day. Perhaps it will allow farmers to reduce fertiliser applications by enabling crops to obtain their own nutrients from the soil or atmosphere. Perhaps plants will be engineered to produce plastics, reducing the dependence of the chemicals industry on fossil fuels, or to produce antibodies for use in medicine.[4] In this chapter, we shall focus on three immediate goals of the genetic engineering of plants, as this will allow us to evaluate better the arguments for and against their development.

Tastier tomatoes

Tomatoes are big business. Sales exceed those of potatoes or lettuce, amounting to over $3.5 billion in 1993. However, it is generally acknowledged that today's tomatoes all too often have a poor flavour and texture.[5] The main reason is that tomatoes are picked before they are ripe. The benefit of this practice is that it allows the tomatoes to be moved from where they are grown to where they are sold before they go soft. Consequently, the tomatoes are less likely to be damaged in transit. The disadvantage, though, stems from the fact that tomato flavour correlates with the amount of time the tomatoes spend on the parent plant. Picking tomatoes when they are still green leads to relatively unflavourful tomatoes. This is one reason why so many people grow their own. Apart from the personal satisfaction of producing your own food, you get tasty tomatoes.

A second problem with picking tomatoes before they go red is that they then have to be treated with ethylene before being sold. Ethylene is a natural plant growth substance and is responsible for the ripening of tomatoes *in situ*. It is supplied to tomatoes that are picked when still green as otherwise they fail to ripen. A final problem with harvesting tomatoes when they are still green is that it is all too easy to pick them when they are still very unripe. These immature green tomatoes taste even less good than the ones that result from tomatoes picked when 'mature green'.

For a variety of reasons, therefore, there are strong incentives to breeding tomatoes that can be picked when red. Such tomatoes should taste better and be firmer. They wouldn't need to be treated with ethylene and might even end up costing less, as the wastage that comes from picking immature green tomatoes would probably be reduced.

Since the mid-1980s, a race has ensued between several companies trying to manufacture and market genetically engineered tomatoes with these characteristics. In 1994, this research reached commercial fruition when Calgene's 'Flavr Savr' tomato went on sale in the USA. Other companies still involved in the genetic engineering of tomatoes include Monsanto and Zeneca Seeds.[6,7]

Two main approaches have been tried in the genetic engineering of tomatoes. One has been to find a bacterial enzyme that prevents the synthesis of ethylene. The thinking here is that since ethylene promotes tomato ripening, genetically engineered plants with low levels of ethylene expression might ripen slowly. This would allow them to be picked when red, because they would remain firm for longer than normal tomatoes – which soon go all squishy if picked when red. An appropriate gene has been isolated from a bacterium called *Pseudomonas* and bred into strains of tomatoes. Research has shown that these tomatoes don't take any longer to ripen on the vine but, when picked, remain firm for longer.[5] As yet, however, this approach has not been commercially advantageous. Instead a quite different, and ingenious, approach has proved successful.

To understand how this other approach works, we need to go back to the principles of protein synthesis described in Chapter 2. The various proteins made by a tomato fruit, including all its enzymes, derive, ultimately, from the DNA in the tomato's chromosomes. This DNA is responsible for the synthesis of messenger RNA – the molecule that carries information from the chromosomes in the nucleus of a cell to the ribosomes in the cytoplasm of the same cell (see pp. 16–19). It is on the ribosomes that messenger RNA is used to make the particular protein coded for by the gene that gave rise to that messenger RNA.

Suppose, though, that the cell could be persuaded to make a molecule that bound to the messenger RNA made by a particular gene. In such a case, the messenger RNA would be unable to synthesise the protein coded for by that gene. It would almost be as if the gene had been excised from the DNA.

The technology to do this now exists and results in the synthesis of so-called 'anti-sense RNA'. In the case of the tomato, the approach taken has been to try to prevent the synthesis of an enzyme called polygalacturonase – PG for short. In a natural tomato, PG synthesis only takes place as the tomato ripens from green to red. Its effect is to soften the fruit by breaking down some of the compounds in the cell walls between the cells of the fruit. From the point of view of the original, wild tomato plant, PG synthesis is highly desirable because it softens the fruit so that it becomes easy for birds to attack. The birds digest the soft fleshy fruit, but the tomato seeds pass unharmed through a bird's digestive system to be dispersed to pastures new.

Preventing PG synthesis doesn't affect ripening, but it does mean that the fruit remains firm – just what is wanted by producers and consumers. It also means that tomato sauces made from the tomatoes are more viscous and so flow less readily. This, too, is a desirable characteristic – runny tomato ketchups are not popular. Indeed, tomato paste manufacturers sometimes heat tomatoes so as to inactivate PG. However, this heating costs money and reduces the flavour.[7]

So for a variety of reasons, there are commercial arguments in favour of preventing PG synthesis. In essence, the anti-sense RNA approach involves the following steps. First, work out the sequence of bases of the DNA that codes for PG. Secondly, in the laboratory make complementary DNA from this DNA. A complementary DNA is one in which each base is the complementary – i.e. matching – base to the base in the original DNA. The first 15 bases in the coding region of the PG gene are ATGGTTATC-CAAAGG. Consequently the corresponding bases of the complementary DNA are TACCAATAGGTTTCC, as A (adenine) binds with T (thymine) and C (cytosine) with G (guanine).

Having made, in the laboratory, the complementary gene to the one responsible for the synthesis of the enzyme PG, the next thing to do is to insert it into the tomato DNA. In reality, there is a bit more to it than that. In particular, you need to insert a gene called a 'promoter' alongside the complementary DNA. The job of the promoter is to make sure that the tomato uses the complementary DNA to make messenger RNA, rather than allowing it to just sit there inactive.

Our story is nearly complete. What was hoped was that the tomato would not only make messenger RNA from the PG gene, but also messenger RNA from the complementary gene. This indeed is exactly what happened. What you find is that the messenger RNA made from the PG gene and the messenger RNA made from the complementary gene are themselves complementary. As before, the first 15 bases in the PG gene are ATGGT-TATCCAAAGG. As a result, the first 15 bases of the messenger RNA made from the PG gene are UACCAAUAGGUUUCC, since messenger RNA has U (uracil) rather than T (thymine) This RNA is called 'sense RNA' since it is the normal messenger RNA. At the same time, the corresponding bases of the messenger RNA made from the complementary gene are AUGGUUAUC-CAAAGG. This RNA is called 'anti-sense RNA' since it is, in a way, opposite in the sequence of its bases to the sense RNA.

Now, here's the elegant part of the story. The two messenger RNAs made fit snugly together. In fact, they fit so tightly together that they don't allow transfer RNAs to bind to them. As a result, neither the sense RNA nor the anti-sense RNA ends up making protein. This is precisely what the plant breeder wants. In the case of the enzyme PG, genetically engineered tomatoes show only about 1% PG activity relative to normal tomatoes.

Are genetically engineered tomatoes genetically engineered?

The point of this, perhaps, rather surprising question – to which the answer must be 'yes' – is to clarify the nature of anti-sense technology such as this. The tomatoes are indeed genetically engineered: they have inserted into them a foreign gene. However, this gene is an artificial one; it doesn't come from another species, or even from a different tomato variety. Instead it is made in the laboratory, the sequence of its bases specifically designed to stop one of the normal tomato genes from having its usual effect.

The nature of anti-sense technology may have regulatory implications too. The UK's Department of Health's Advisory Committee on Novel Foods and Processes has put forward guidelines on the assessment of novel foods and processes.[8] Tomatoes genetically engineered through anti-sense technology have successfully passed the necessary tests, all of which relate to safety for a consumer ingesting such tomatoes. The same committee, in the same document, then went on to address the issue of labelling:

> The Food Advisory Committee has been considering the role
> which genetic engineering is beginning to play in food
> production and has been concerned to identify those uses
> of the technology which could present moral and ethical
> concerns for some consumers. The Committee has concluded
> that there are some uses of the technology in the production
> of food which would generate such concerns because of the
> possible presence of genetically modified organisms (GMOs)
> in the final food. The Committee felt that labelling was not
> the answer to these concerns except where the presence or
> use of GMOs could be considered to alter materially the
> nature of the food.[9]

This conclusion is based on the assumption that the only moral and ethical concerns that consumers have relate to the nature of the food product, rather than to the processes by which it has been produced. As a result of these guidelines, a number of genetically engineered foods do not, at the time of writing, require labelling in the UK. These foods include genetically engineered chymosin (see pp. 118–120) and genetically modified baker's yeast. It is likely that most people in the UK, though doubtless unaware of the fact, have eaten food made as a result of genetic engineering. It is also the case in the UK that milk obtained from cows injected with genetically engineered BST has been allowed to mix with the rest of the milk supply. It is likely that tomatoes genetically engineered by anti-sense technology would similarly not be required to be labelled distinctively. The ethical issues relating to labelling were examined in the previous chapter.

To keep all this in proportion, it needs to be realised that the modern tomato, *Lycopersicum esculentum*, has already had many features bred into it by hybridisations, through traditional techniques of plant breeding, between a number of different *Lycopersicum* species (Figure 6.1). It might, therefore, be argued that the genetic integrity of the tomato has already, through conventional plant breeding, been somewhat violated.

A slight complication is that when researchers genetically engineer a species, they often add a 'marker' gene that makes the organism immune to a particular antibiotic. The reason for this is that it makes it easier in the laboratory to see whether the genetic engineering has worked. In the case of tomatoes, for example, you simply have to see if the young tomato seedling is unaffected by the presence of an antibiotic such as kanamycin or neomycin, rather than waiting to see if the fruits the adult tomato eventually produces take longer to go rotten.

The possibility has been raised, not least by the UK's Department of Health's Advisory Committee on Novel Foods and Processes in its 1994 report, that there is a chance, albeit a very small one, that when large amounts of foods containing these antibiotic marker genes begin to be consumed, the gene might move to disease-causing microorganisms in the gut and so make them resistant to the antibiotic too.

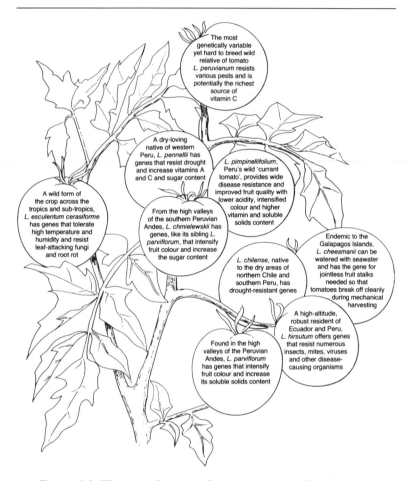

Figure 6.1. The contributions of various species of *Lycopersicum* to the modern tomato. These contributions have not been the result of genetic engineering, but of various more traditional techniques of plant breeding.

Most experts suspect that the chances of this happening are not great.[10] Even if it does happen, the consequences are unlikely to be desperately serious as there are many different types of antibiotic and any one marker gene only conveys resistance to one of them. Nevertheless, the existence of a finite risk may slow the regulatory approval of genetically engineered tomatoes and other foods in the UK. In the long run, the most likely solution is for companies to use other, less controversial markers. Technically this is possible, though less easy.[11]

One other objection to these tomatoes should be mentioned. It has been argued that their longer shelf-life means that they lose more of their vitamin C content before being eaten. It is, indeed, the case that vitamin C breaks down over time. However, even if genetically engineered tomatoes end up, on average, with lower levels of available vitamin C, it is difficult to imagine that the effect is likely to be of great nutritional significance.

Herbicide-resistant crops

Herbicides are chemicals that are used to kill plants. Unwanted plants are generally referred to as weeds and there can be many reasons why weeds are considered undesirable. The most frequent reason is that weeds compete with a food crop for nutrients, water or space. Weeds can be dealt with in a number of ways: they can be removed by hand, for example. In many countries, herbicides are extensively used to control weeds. Indeed, approximately 90–95% of the area of land used to grow crops in Europe and the USA is treated with herbicides each year.

An ideal herbicide would have the following characteristics (in no particular order):

- It would kill a wide variety of weeds
- It would have no harmful effects on the growth of the crop
- It would be harmless to other organisms, including animals, humans, fungi and soil bacteria
- It would be biodegradable, quickly breaking down to harmless chemicals, thus avoiding any build-up of toxic residues in the soil
- It would have low leaching potential through the soil profile, meaning that it would be unlikely to reach ground water[12]
- It would be cheap
- It would be easy to apply.

Why genetically engineer herbicide-tolerant crops?

At present a large number of different herbicides are available commercially. Not surprisingly these differ considerably in their environmental impact, i.e. their biodegradability and effects on non-target species such as insects, birds and mammals. An argument from those who favour the widespread use of genetically engineered herbicide-tolerant crops is that by genetically engineering the crop to be resistant to environmentally acceptable herbicides the overall impact of herbicide use on the environment will be lessened. For example, the herbicide paraquat is widely used by farmers and others who wish to control weeds, because it acts against a broad spectrum of weeds. It breaks down quite rapidly and has a low leaching potential. However, it is toxic to a wide range of animals. Another commonly used herbicide has atrazine as its active ingredient. Atrazine has a low toxicity, but lasts for a long time in the soil before being broken down. In some places it has become a problem through reaching ground water.

Genetic engineering holds out the hope that instead of researchers starting with desirable crops and then finding herbicides that kill the weeds infesting those crops without harming the crops, they could start with the most desirable herbicides and then genetically alter the crops so that they, unlike their weeds, are unaffected by the chemicals. Two of the most suitable herbicides are glyphosate (better known to many by its trade name Roundup) and glufosinate (Basta or Challenge).

But who precisely is expected to benefit from herbicide-tolerant crops? Benefits to the farmer, the environment and the consumer have been suggested.[12]

The farmer is predicted to benefit for a number of reasons. First of all because glyphosate, and other herbicides to which crop tolerance is being sought, are generally less expensive to purchase or easier to apply than their alternatives. Secondly, inducing herbicide tolerance in a crop increases, it is maintained, a farmer's flexibility because it means that an extra herbicide may now be available. Any existing herbicides can still be used because the fact that a crop has been genetically engineered to be tolerant to a particular herbicide does not mean that *that* herbicide – or indeed any herbicide – has to be used. Thirdly, herbicides such as glyphosate are particularly effective. Their use should, therefore, lead to a greater range of weeds being controlled, resulting in higher crop yields.

Benefits to the environment are anticipated because the herbicides to which crop tolerance is being developed generally break down faster to non-toxic products in the soil and are less likely to leach into groundwater. In addition, they are often active in smaller amounts, meaning that the total mass of chemicals applied to a crop should be less. Finally, some of these herbicides reduce the need for pre-emergence application (when the herbicide is applied before the seed germinates). The advantage of this is that post-emergence application can reduce the risk of erosion of fragile soils.

Benefits to the consumer are envisaged on two fronts (Figure 6.2). First of all the technology, through increasing crop yields and requiring farmers to spend less on herbicides, should decrease food prices. Secondly, the technology, as already suggested, should lead to lower rates of herbicide application, the use of less toxic chemicals and a decreased risk to domestic water supplies.

"It's the best thing since, well, er....sliced bread !"

Figure 6.2. Some people may have unrealistic expectations about the extent to which genetic engineering will improve crops.

Arguments against the genetic engineering of herbicide-tolerant crops

A number of reasons have been proposed as to why the genetic engineering of herbicide-resistant crops is not a good idea.[13] Two of these arguments apply to any genetic engineering of plants: namely that this increases the chances of the crop invading and then damaging natural ecosystems, and that it increases the chance of the genes inserted into the crop escaping into weeds. We will examine these two arguments in detail later in this chapter when we consider the breeding of disease-resistant crops (p. 147). A further argument – that genetic engineering is intrinsically wrong, for example because it nearly always involves the transgression of species boundaries – was considered on pp. 61–3. This is, of course, an argument against almost any genetic engineering, not just against the genetic engineering of plants.

Here we examine specific objections to the breeding of herbicide resistance in crops through genetic engineering. It must be noted, however, that these objections are not, in essence, objections just to genetic engineering. They would hold even if conventional crop breeding were used to bring about herbicide resistance. In reality, though, herbicide-resistant crops are more likely to be developed through genetic engineering than through conventional plant breeding.

At first sight, the arguments advanced in favour of the genetic engineering of crop plants to promote herbicide tolerance may sound impressive. There are suggested benefits to the farmer, the consumer and the environment in terms of cost and of safety. However, there are those who counter these arguments, as we shall see. It should be noted that the disagreements are mainly about the actual outcomes of the technology. In other words, they are arguments about facts and about consequences, not about values or rights. For this reason, there are some grounds to hope that the areas of disagreement may reduce over time, given good quality empirical data and informed debate.

One of the main arguments against the new technology is that it will concentrate power in the hands of a few agrochemical companies. Farmers, it is maintained, will become locked into a dependency on these companies with little they can do about significant hikes in herbicide and seed prices. At its extreme, this fear belongs to the conspiracy-theory genre and, to caricature somewhat, envisages powerless farmers forced to pay ever greater amounts to anonymous international companies who profit both from the cost of the crop seed and from the cost of the herbicides used to spray them. The counter to this argument is to maintain that farmers have choices. In particular, they have a choice when they decide what seed to purchase and they have a choice when they decide whether or not to spray. But it can be argued that farmers, although not physically forced into purchasing herbicide-tolerant seed and then using the accompanying herbicide, will end up paying more than they would otherwise have done.

Essentially this is an argument about the free market, about monopolies and about price-fixing. Those in favour of the development of herbicide tolerance in crops may emphasise the farmer's autonomy, the benefit of a free market system and the existing power of the regulatory authorities to ensure that unfair monopoly practices do not result. Those against the development of herbicide tolerance in crops may argue that this rosy scenario is politically naive and fails to correspond to the reality, both in Western and developing countries, of the ever-increasing amounts spent by farmers on herbicides.

A second argument against the development of herbicide-resistant crops is that the introduction of herbicide-tolerant crops will lead to the use of more herbicides. There is no need to spell out the two opposing views in great detail. Those in favour of the new technology assert that herbicide use is already widespread and that we should be looking at the impact of the herbicides used, not merely at the amounts used, which, in any case, should decrease given genetic engineering. Those against the new technology point out that herbicide use is not at present ubiquitous, particularly outside Europe and the USA, and that it stands to reason that agrochemical companies are investing millions of dollars in all this because they hope that herbicide use will increase.

To give a final example, someone opposed to the use of genetic engineering to engender herbicide tolerance might, at best, smile wryly at the announcement of the successful genetic engineering of herbicide resistance in *Agrostis palustris*.[14] This plant is a type of turf grass used for golf greens and high quality intensively cultured fairways, tees, lawns and similar turfs. Why on earth, one might ask, are we bothering to develop genetically engineered herbicide-tolerant grasses for golf courses? Surely we should be using fewer, not more herbicides on such places? Doesn't this sort of development show the shallowness of an industry that with its public persona claims to be all about enhancing food production and so saving people from misery and starvation but in reality is concerned merely to grab every available dollar of revenue even if this means using large volumes of herbicides to create artificially manicured ecological deserts only to be enjoyed by people with above-average incomes?

However, someone in favour of such a development might point out that in a free society people should surely be allowed to keep weeds off golf courses if they want to. If members of such golf courses are prepared to pay the extra fees needed to pay for the seeds, the herbicides and their application, that is their choice. What is the alternative? To ban all of us from buying herbicides and insecticides – even slug pellets – for use in our own gardens? Anyway, as revealed in a paper titled 'Mortality study among golf course superintendents', at present the most commonly used herbicide on golf courses in the USA is the phenoxy-herbicide 2,4-D.[15] Unlike 2,4-D, the herbicide for which resistance has now been genetically engineered in *Agrostis palustris* (Finale; Hoechst Roussel) degrades rapidly in the soil into naturally occurring compounds and has no soil residual activity. The technology is, therefore, good for the environment – and for golf superintendents.

Pest-resistant crops

The prospect that genetic engineering might greatly increase crop yields is perhaps the single strongest argument in favour of the genetic engineering of plants. Those in favour of the genetic engineering of crops argue that increasing human population sizes mean that more food is needed.[16] The day may come when genetic engineering will allow increased crop yields through the development of crops that can grow in more arid or more salty habitats than at present, or that can fix nitrogen from the air: in essence making their own fertiliser.[17] In the immediate future, though, genetic engineering is perhaps most likely to increase crop yields by making plants more resistant to pests. For this reason, we shall concentrate on this aim of crop biotechnology. We use the word 'pest' in its broadest sense to include any organism that reduces crop yield. The term, therefore, includes insects and other animals that eat a crop either as it grows or after it is harvested, fungi and viruses that cause plant diseases and weeds that compete with a crop.

How important are crop pests?

There is no doubt that crop pests very significantly depress yields and cost enormous amounts of money. For instance, the gall midge is a pest of rice and causes worldwide losses in rice yields equivalent to approximately $550 million at 1994 prices.[18] In the USA in 1990, some $350 million was spent on pesticides merely designed to kill pests of cotton (an average of over $30 per acre of the crop). Yet these pests still caused some $270 million of damage.[19] The European corn borer (*Ostrinia nubilalis*) is a major pest of maize in North America and Europe. Crop losses in the USA alone are worth approximately $500 million a year at 1992 prices.[20] In many developing countries viral infections reduce the production of root and tuber crops by between 20 and 80%.[21] Worldwide, around a third of all potential crop production is lost through pests.

Possible benefits of genetically engineered pest resistance in crops

Advocates of the genetic engineering of pest resistance in crops maintain that it will lead to an increase in yields, a decrease in the use of pesticides and a decrease in the price the consumer pays for crop products. In the case of cotton, for example, it is hoped that the insertion of a gene naturally found in bacteria will reduce insecticide applications for lepidopteran pests (caterpillars) by more than 50%. In the case of maize, field trials of plants

Figure 6.3. Wishful thinking by certain crop biotechnologists.

genetically engineered to be resistant to the European corn borer have proved outstandingly successful: plants genetically engineered with an insecticide of bacterial origin have shown only a small fraction of the damage suffered by control plants.[22]

Although it is still early days, there are genuine reasons to expect genetic engineering to increase crops yields in plants as diverse as cotton,[17] maize,[22] rice,[23] and apricots,[24] to give just four examples (Figure 6.3).

Nor should it be thought that genetic engineering is likely to benefit only Western countries. It is true that some of the uses of genetic engineering in plants may cause anyone with a social conscience to wince – the first field trials of genetically engineered plants were carried out (in China) on tobacco, while intensive research is under way to produce genetically engineered roses with blue flowers and Christmas trees that don't shed their needles.[25] Nevertheless, genetic engineering can be expected to bring considerable benefits to developing countries.

Consider, for instance, cassava.[26] Some 600 to 800 million people in the tropics rely on cassava for food. It is particularly valuable in dry regions of Africa because it can grow without rain for as long as four months. Yet cassava suffers from several viral diseases and insect pests. The African cassava mosaic virus on its own reduces yields by around 30 to 80%. In 1991, Roger Beachy established the International Laboratory for Tropical Agricultural Biotechnology. This laboratory is dedicated entirely to the improvement of crops of developing countries. Projects have focused on cassava, rice, sugar cane, sweet potato and tomato. Beachy has succeeded in genetically engineering resistance to viruses in a number of plant species, including rice, and there is a realistic expectation that genetic engineering should be able to help protect cassava against viruses and other pests.

Nor should it be thought that advances in the developing world (itself a phrase felt by many to be patronising, but used here for want of a really suitable term) must rely on hand-me-downs from philanthropic or conscience-stricken Western scientists. For example, the lead in the application of genetic engineering to the protection of rice against the gall midge is being taken in India. In 1994, a breakthrough in this field came when researchers at the International Centre for Genetic Engineering and Biotechnology in Delhi, headed by Krishna Kumar Tiwari, discovered a way of drastically cutting the time taken to identify resistant varieties.[18]

Are there ecological risks associated with genetically engineered pest resistance in crops?

So genetic engineering of pest resistance in crops offers great benefits. But what are the ecological risks? Two main ones have been suggested. First that the technique will increase the chances of the crop invading and then damaging natural ecosystems, and secondly that it will increase the chance of the genes inserted into the crop escaping into weeds.

The dangers of crop invasion

At first sight, the notion that genetic engineering might cause crops to become weeds, resulting in the altered crops invading and then damaging the environment, may sound far fetched. After all, most of us are used to seeing crops such as wheat or maize or vines growing over huge areas without any escapes into the wild. However, there are a number of reasons why this threat should be taken seriously.

For one thing, introduced plants do sometimes run amok. We have already quoted (p. 111) the example of the prickly pear cactus introduced to Australia at the beginning of the twentieth century. By 1910, the cactus was spreading at a rate of over 300 000 hectares a year. Only the introduction of eggs of the moth *Cactoblastis cactorum* from Argentina saved the day. The caterpillars fed on the prickly pear cacti and so brought the problem under control.

It might be objected that the prickly pear cactus was introduced as an ornamental plant (for its supposed beauty) not as a commercial crop. However, there are examples of plants introduced as food and forage crops later becoming weeds. For example, Johnson grass (*Sorghum halepense*) was brought to the USA in the early nineteenth century as a perennial forage grass but is now considered one of the world's ten worst weeds.[27] Other plants introduced to the USA as food and forage crops that have subsequently become serious weeds include artichoke thistle (which has invaded rangelands), cogongrass (which has invaded southern farms), crabgrass (which has invaded pastures and croplands) and reed canary grass (which has invaded canals and ditch banks).[27] Nor is the problem confined to the USA. A number of other countries have similar problems.[28]

So how great is the problem? A rough guide is given by the 'ten–ten–ten' rule.[29] This applies to all groups of organisms, including plants. When applied to plants it states that approximately 10% of the species grown deliberately or imported accidentally become *introduced* (or casual), that is the individual plants can live independently without being intentionally planted. Approximately 10% of these introduced species go on to become *established* – meaning that they have permanent populations. Finally, approximately 10% of these established species go on to become *weeds* – meaning that they have serious economic consequences or non-trivially damage natural ecosystems. In reality, however, crop plants are more likely than this to become weeds.[30] This may be because nearly all food and fibre crops have close relatives that are regarded as weeds somewhere in the world.

Is there any way then that we can predict in advance which plants will become weeds? Classic work by Baker in the 1960s and 1970s led to the identification of a list of plant features characteristic of weeds.[31] These include such things as:

- High number of seeds produced under favourable conditions
- Some seed output even under unfavourable conditions
- Seed production begins after only a short period of vegetative growth
- Seed can survive in the soil for many years
- Rapid seedling growth and establishment
- Plant able to tolerate considerable variation in the physical environment (e.g. climate, different soil types)
- Long-distance dispersal possible
- Good powers of vegetative reproduction and the ability to regenerate when divided into fragments
- Strong competitive ability
- Seed production following both cross-pollination (between two separate plants) or self-pollination.

Fortunately no one weed possesses all these characteristics. Unfortunately ecology is not yet a very predictive science. Attempts to predict which plants become weeds have not been successful. As Mark Williamson, Professor in the Biology Department at the University of York, concludes (p. 75): 'Baker characters have no predictive value for weediness'.[30]

It might be objected that even though we may not be able to predict with any great accuracy whether a particular plant will become a weed, that isn't the central issue. What we are concerned with is whether introducing one, or at most a few, new genes into a plant is likely to cause it to become a weed. This may be the right question. Unfortunately the answer is still unclear. On the one hand, it is easy to find people with impressive academic credentials who argue that the likelihood of the addition of only a handful of genes to a plant suddenly turning it into a pernicious weed is negligible.[32] On the other hand, there are others, with equally impressive academic credentials, who argue that small genetic changes can indeed cause very significant ecological changes.[33]

Given this genuine controversy, there is little one can do but suggest that we continue to proceed with caution on a case-by-case basis, so that the chance of genetic engineering producing new weeds can carefully be examined.[34] The biggest study carried out to date to see whether genetic engineering increases the chances of a crop invading other areas was carried out by Michael Crawley and his colleagues on oilseed rape in the UK.[35] Crawley worked in three contrasting parts of the country – Cornwall, Berkshire and Sutherland. In each of these three places, four different sites were chosen, and at each site two different genetically engineered and one control (i.e. non-genetically engineered) line were planted. The whole experiment was run for three years. The results are easy to summarise. The genetically engineered lines were no more invasive – if anything, less invasive – than the control line.

Crawley's experiment, which cost £1 million, has been widely cited by those in favour of the genetic engineering of crops. However, it has been criticised. Williamson, the Professor of Biology at York University whose work we have already cited, has been quoted as saying 'The experiments asked "is oilseed rape going to become a pest?" in habitats where we already know the answer is "no"'.[36] Similar concerns about Crawley's work have been expressed by others.[37]

Further, there are those who suspect that even the apparently sensible approach of proceeding on a case-by-case basis is doomed. There can, for instance, be surprisingly large delays between the introduction of a plant and its emergence as a weed. For example, proso millet became a weed only after more than 200 years of cultivation.[38]

Given all this, a more profitable way forward, though not one which we are aware has so far been taken, might be to accept that there is a genuine likelihood that one or more genetically engineered crops will become weeds, and then to figure out ways of minimising the damage, both ecological and economic. Such an approach seems sensible given that the public has become all too used to scientists and industrialists – whether in the nuclear, chemical, pharmaceutical or agricultural industries – maintaining at one time that their products will never cause problems, only for us all to find out, a few years later, that life rarely holds such absolute guarantees.

The dangers of genes inserted into a crop escaping into weeds

A second danger is not that genetically engineered crops will themselves become weeds, but that the genes inserted into them will escape into other plants that then become weeds, damaging natural ecosystems or other areas. It is difficult to generalise about the magnitude of this risk. At one extreme are plants such as maize and soybeans in the USA. These have no sexually compatible wild relatives in the USA. Even those who consider that the genetic engineering of crop plants poses considerable ecological risks admit that for these crops 'the likelihood of gene flow into wild populations as a result of use in the United States is virtually nil' (p. 29).[27]

However, there are other crops where this is much more of a risk. For example, crop beet (*Beta vulgaris*) readily hybridises in the UK with wild beet (*Beta maritima*) and in California with another species of wild beet (*Beta macrocarpa*). Weed beet has arisen from crosses of *Beta vulgaris* with the other two beets. Weed beet is commonly controlled by applications of the herbicide glyphosate. At present there is a programme to produce crop beet genetically engineered to be resistant to glyphosate. There is, therefore, a non-trivial risk that glyphosate resistance may spread to weed beet populations.[39]

Nor is this an isolated problem. In the USA, it has been argued that the majority of the economically important crops have wild relatives with which they are sexually compatible.[27] The problem is also considered serious in the UK[39] and in the Netherlands.[40]

Nevertheless, the data we have discussed so far do not relate to the specific risks associated with the movement of genes from genetically engineered crops into other plants. Research in this area is still in its infancy. Philip Dale at the John Innes Centre in Norwich carried out the first British field release of a genetically engineered plant, a potato.[41] More recently, he has looked at the extent to which gene flow can occur between potatoes and two native weeds in the same genus, black nightshade (*Solanum nigrum*) and woody nightshade (*Solanum dulcamara*). Encouragingly, Dale found no evidence for cross-pollination between the crop and the two weeds. However, both this experiment and the one we cited earlier carried out by Crawley on oilseed rape[35] can be criticised on the grounds that they only investigated crops that had been genetically engineered with marker genes – antibiotic resistance in the potato and antibiotic and herbicide resistance in the oilseed rape. The conclusions from these experiments would be more convincing if the transferred genes had conferred some definite ecological advantage such as increased frost tolerance.

Other arguments against the genetic engineering of pest resistance in plants

There are other potential dangers of the genetic engineering of pest resistance in plants, though it must be emphasised that all of these claims about the dangers of genetic engineering are controversial.[42] For example, Henry Miller, Director of the Office of Biotechnology at the US Food and Drug Administration, has argued that: 'The U.S. Department of Agriculture (Washington, D.C.), the Environmental Protection Agency (Washington, D.C.) and the European Community (Brussels, Belgium) have built huge, expensive, and gratuitous biotechnology regulatory empires preoccupied with negligible-risk activities, and have succeeded in protecting consumers only from enjoying the benefits of the new technology'.[43]

So what are these other potential dangers to the environment, to the economy or to people's health? First of all, plants genetically engineered with virus particles may lead to the evolution of new viruses, some of which might pose dangers for existing crops.[44] Secondly, plants genetically engineered to produce toxic chemicals – thus deterring the pests that feed on them – might poison native animals, fungi or bacteria.[27] Thirdly, the widespread use of genetically engineered plants in developing countries may even further accelerate the current loss of genetic diversity that is happening as the number of natural varieties (land races) declines through irreversible extinctions.[27] Fourthly, moving genes from one plant species to another may increase the risk of putting allergens into food.[45]

Then, there are at least three further objections raised to the genetic engineering of pest resistance in plants: first, that so-called 'ethically sensitive' genes may be inserted into common crop species; secondly, that the technology may end up exploiting farmers and others in developing countries; thirdly, that we simply don't need this work.

The insertion of ethically sensitive genes into plants

Consider the example of potatoes growing high in the Andes and elsewhere. For many years, the International Potato Centre has been trying, for obvious reasons, to breed frost-resistant potato varieties. Conventional plant breeding has had some success, but progress has been slow. Recently a breakthrough has come about through genetic engineering.[46] Researchers have succeeded in moving one of the genes found in a particular species of fish – the flounder *Pseudopleuronectes americanus* – into potatoes. Why – one might ask. The reason is simple. The gene makes an anti-freeze protein that allows the flounder to survive and flourish at sub-zero temperatures. By 1994, work in Bolivian glasshouses had shown that the genetic engineering was successful. The next stage is to carry out large-scale field trials of the new potatoes.

However, this work does raise a moral issue. Putting it bluntly: is a potato with a flounder gene still a plant? Or is it part-plant, part-animal? To a vegetarian, even more to a vegan, the question is not merely an academic one. Suppose, to put it at its most extreme, most potato varieties in ten years' time carry the flounder gene; will potato salads, baked potatoes and French fries be things of the past for vegetarians?

Might the genetic engineering of plants end up exploiting farmers and others in developing countries?

It has been argued that genetic engineering, with its attendant patents, is unfair to farmers and others in developing countries. Indeed in most countries of the world farmers keep back some of their harvest as next year's seed. Even in the UK, only 70% of the seed sown is purchased: around 30% comes from saved seed produced on the same farm.[47] Increasingly this practice is being restricted by the companies that supply the seed. From the point of view of such a company, the royalty it gets from the sale of seeds are its primary source of income. Consider, though, the broader arguments of the *Seed Satyagraha*, the fight for truth (or seed movement), in India.[48] The seed movement argues both that the patenting of life is ethically unacceptable in principle and that it unfairly discriminates against farmers in developing countries. To give just one example, Vandana Shiva, the leader of the movement, has protested against the patenting by a number of multinational corporations of extracts obtained from neem (*Azadirachta indica*). This is a tree widely grown in India. Ironically, its formal botanical name is derived from the Persian *azad darakht*, which means 'the free tree'.

> The *Neem* is a beautiful tree. It really looks regal. It grows best in arid zones. The poorest of homes will have a *Neem* in the backyard. The *Neem* has terrific anti-malarial properties – it doesn't allow mosquitoes to come near; it doesn't kill mosquitoes, it numbs them and keeps them away. It has been used by our mothers and grandmothers; they used the dry leaves in clothing, so that silk and wool did not get eaten by worms. *Neem* leaves have been used in storing grain, so that, again, bugs don't get to the grain. *Neem* is a reliable skin treatment for all kinds of infection. My own little boy used to pick up infections all the time. The only thing that would work was the *Neem* oil massage I gave him. It is now being found to be very effective as a contraceptive. The *Neem* is a sacred tree in India. It is the olive of India. It's always been known that if you use the *Neem* twigs as a toothbrush, you never get any kind of tooth decay; but there's a US company that now has a patent on its dental care properties. There's a company that has a patent on its skin care properties; and, of course, you have about ten companies which have patents on its biopesticide properties. So every aspect of *Neem* that has been known in India is being treated as an innovation of a Western corporation. (p. 38)[48]

A report commissioned by the UN concluded in 1994 that 'bio-piracy' is cheating developing countries and their indigenous peoples of some \$5.3 billion a year.[49] Developing countries do, of course, have many times more species (and hence types of gene) than do Western countries. Almost all the world's crops were first developed in today's developing countries.

Is the genetic engineering of plants for pest resistance necessary?

Finally, it is worth outlining some factors that might help one to question the need for the genetic engineering of plants for pest resistance. First, there is little doubt that insects and other pests will evolve virulence and so be able to attack these plants.[50] This, of course, does not mean that the technology is a waste of time. What it does mean is that the fact that pests destroy around a third of today's crops tells us little about the extent to which genetic engineering will be able to reduce this figure.

Secondly, one should always be cautious about someone telling one that something is necessary. If one really wanted to, there is little doubt one could do without genetic engineering in general and the genetic engineering of pest resistance in plants in particular. For one thing, there are major advances still being made in conventional plant breeding both in terms of pest resistance and yield enhancement.[51] Then – to take an extreme position – suppose that the genetic engineering of pest resistance in plants allows the world human population size to stabilise at 12 billion rather than 10 billion people. Which of us can confidently forecast which is the better scenario both in terms of human quality of life and in terms of our effect on the other species with which we share this planet? Such arguments caution us against uncritically accepting the need for genetic engineering in plants, without, of course, meaning that this need can lightly be dismissed.

CAN PLANTS RAISE ETHICAL PROBLEMS?

The discussions earlier in this chapter suggest that the answer to this question is yes. The genetic engineering of plants causes some people to feel moral concern for a variety of reasons, and we now need to look at some of these in a little more detail to examine their ethical basis.

Some of these concerns are of a general kind, applying to all forms of genetic engineering, and these have already been considered in earlier chapters. Intrinsic objections to the genetic engineering of plants, for example, are seldom voiced but are theoretically possible if a person felt that it was wrong in itself or 'unnatural' to tamper in this way with the plant kingdom. Objections of this kind can be imagined in the case of such genetic innovations as blue roses or carnations, though the grounds for these objections might be more aesthetic than moral (e.g. 'that's not how roses should look' rather than 'it's wrong to do that to roses').

Intrinsic objections have been fully dealt with in earlier chapters, however, and plants do not seem to raise any distinctively new issues of this kind. It is mainly extrinsic concerns, therefore, that need examining in this chapter, and even here we find that in some cases the main points have already been made. The ethical significance of risk and safety, for example, has been analysed in Chapter 3 and illustrated in Chapter 5, and no more needs to be said here, except to re-emphasise the point that extreme caution may not always turn out to be the 'safest' option in the long run, if it deprives us of some agricultural development that may be desperately needed at some future time.

The main extrinsic concerns, however, that are raised by the genetic engineering of plants and which have been referred to already in this chapter are socio-economic ones that centre around the ethical principle of fairness. Is it fair that small farmers and vulnerable economies may be the losers from this technology? Is it fair to allow seeds to be patented? Is it fair to use and profit from genetic resources taken from poorer countries? How does the ethical arithmetic work out here?

Exploitation and new technologies

As with the concerns about risk, the issues at stake here are in one sense at least clearcut and uncontroversial. Just as no one is likely to defend irresponsible and potentially catastrophic risk-taking, similarly no one is going to declare support for 'exploitation', 'misappropriation' or 'unfair control'. Like sin, these are things that everyone is 'agin'. The problem, when such language is introduced, is to decide what *counts* as 'exploitation', 'misappropriation', etc. and to judge whether the likely consequences of genetically engineering plants fall within these categories. A detailed, technical assessment of those consequences is beyond the scope of this book, but the following general considerations need to be noted in any analysis of these concerns:

1. History has demonstrated how all new technologies inevitably have far-reaching socio-economic effects. The genetic engineering of plants cannot be singled out as the sole target for moral censure on these grounds, any more than can information technology or the development of the steam engine. Accordingly, as the American philosopher Gary Comstock puts it: 'The argument here cannot be that the technology will put *some* workers out of business; if we were to object to inventions on those grounds we would have had to oppose railroads, electricity and electronic printing presses'.[52]

2. All new technologies tend to benefit *initially* the 'developed' countries, because they have more resources and expertise available for research and development. But the fact that penicillin, for instance, was used to treat Western patients before it became available worldwide is hardly a moral argument against the development of life-saving drugs; though clearly there is a moral obligation upon those benefitting from a new technology to extend those benefits as widely and as speedily as is practicable.

3. The possibility of undesirable effects resulting from an activity does not mean that that activity should automatically be morally condemned and banned. Chimney-sweeping, for example, once led to young children being forced up chimneys at the risk of life

and limb, but the appropriate moral pressure in this case was for *political, legislative* action to prevent these consequences and not for a ban on the sweeping of chimneys (in itself a socially and environmentally desirable activity). Similarly with the genetic engineering of plants, a recent UK report concludes: 'There is no reason why the authorities should be indifferent to the reasonable apprehension . . . that the new biotechnology will bring greater relative advantages to rich farmers and rich countries than to the poor. The problem is not insoluble'.[53]

The problem of patenting

Some of the above points are also relevant to the particular issue of patenting. There are points to be made on both sides here. The patenting of a new product or process by definition places certain restrictions upon others who wish to use that product or process. They may judge the restriction to be 'unfair', but why should this alleged 'unfairness' be more objectionable in the case of a new strain of seed than of a new piece of farm machinery? Despite the charge of 'unfairness', there is on the face of it a straightforward ethical justification for the use of patents in general *in terms of* fairness and justice. If I spend years of my life developing a revolutionary new product that may benefit millions of people, do I not deserve some recompense for my work and some protection against others who might otherwise capitalise upon my efforts? The potential unfairness of denying this protection to biological 'inventions' is illustrated in this lament by the American plant-breeder Luther Burbank in the 1920s *before* any such protection was available in the USA.

> A man can patent a mousetrap or copyright a nasty song, but if he gives the world a new fruit that will add millions to the value of the Earth's annual harvest he will be fortunate if he is rewarded by so much as having his name connected with the result . . . I would hesitate to advise a young man . . . to adopt

plant breeding as a life work until [Congress] takes some
action to protect his unquestioned right to some benefit
from his achievements.[54]

Many of the moral concerns about patenting appear to stem
from a confused view of what a patent in fact is. A recent
theological discussion of the patenting issue states the position
clearly:

> To what extent does patenting imply ownership and is
> ownership of living matter a new and unacceptable human
> domination?
>
> In one sense the difficulty is based on a false premise: a
> patent is not a claim to ownership. A patent is technically the
> right to retain or obtain royalties from others who use the
> invention covered by the patent commercially. The grant of
> a patent does not give the owner of the patent an automatic
> right to do what is covered by the patent. It is in fact a
> negative right; it is the right to exclude others from doing
> something, not a positive right to do something oneself.
>
> In another sense an inventor claims ownership, but it is
> ownership of her or his invention, and the patent is public
> recognition of that. The invention is her or his own. But
> ownership of matter, organic or inorganic, does not
> necessarily follow from this.[55]

Patents, then, are a licence for commercial protection, not a
moral right of ownership. Indeed, ownership is in one important
respect *relinquished* when a patent is granted, for information about
the invention is thereby made public and available for others to use,
provided that they make no commercial gain from it. In this way
patents can, as a result of such public disclosure, facilitate research
and the free flow of information rather than impede it. Admittedly,
the existence of patents can delay the publication of scientific
research and other useful information while the patent is being
applied for. However, this delay is typically of the order of only a
few months.

The emotive charge of 'patenting and owning life' loses much of
its force when it is realised that the 'ownership' involved is of the

invention itself and not of 'matter, organic or inorganic'. In the case of plants, a patent implies ownership not of a seed or a whole plant but of the invention of the 'genetic kit' that produces a particular attribute of that crop. In any case, the logic behind moral indignation about 'owning life-forms' seems far from clear in view of the fact that we happily talk about owning cats and dogs and orchids and orchards without arousing moral outrage.

The claim that genetic resources are the 'common heritage of mankind' and should thus be free for all to use and benefit from also stands in need of closer analysis. Genetic (and indeed all natural) resources are certainly 'free' in the sense that they are present in the natural world without a price label attached, but that does not necessarily mean that they are 'free' for anybody to take and use. There might perhaps be oil beneath the waters of Loch Lomond, but I am not 'free' to set up my drilling rig there to extract it, and even if I were, the operation would be far from 'free' in terms of financial commitment. Few natural resources, therefore, are 'free' in the sense that wild blackberries are. Controls and licences are needed to regulate their extraction, and financial investment is needed to pay for the costs. It may be misleading, then, to imply that genetic resources are like fruits of the forest, free for all to pluck and eat.

The genetic engineering of plants does not, therefore, seem to possess any distinctive features that make it particularly vulnerable to moral concerns about patents and 'free' natural resources. These concerns would need to be shown to be applicable to *all* cases of patenting and to the use of *all* natural resources, but it is difficult to see on ethical grounds what is wrong with patenting a screwdriver or controlling the development of the North Sea oil fields.

Weighing the costs and benefits

No one can guarantee that there will be no undesirable socio-economic effects of genetically engineered crops, and there is likely to be a need for international, political and economic action to prevent or minimise these by showing sensitivity to the problems of the most vulnerable individuals and communities.

But if the possibility of undesirable socio-economic effects is to be added to the possibility of risk, do not these factors constitute a strong argument against the genetic engineering of plants, despite the general considerations mentioned above? This brings us back to the fundamental issue of the value judgments that have to be made about the acceptability and justifiability of the possible costs of the new technology. Those value judgments, however, need to take into account *another* set of factors concerning the potential *benefits* of genetically engineering plants. These have been discussed in the first part of this chapter and somehow have to be weighed against the potential costs.

Even the fiercest critics of the technology find it difficult to deny these substantial, potential benefits, as Jeremy Rifkin illustrates in this extract from a lecture: 'Now people said to me, Jeremy, my God do we want to turn a deaf ear to a technology that could create new plants and animals to feed a hungry world? We wouldn't want to say no to that'.[56] The usual strategy of such critics, however, is to emphasise the potential costs and risks rather than the benefits, without always making explicit the justification for this value judgment, and to suggest that the benefits are of less moral significance than is commonly supposed (e.g. on the grounds that they are more likely to accrue to Western capitalists than to the developing world's farmers.)

Any estimation of the potential benefits will again involve prediction and technical assessment; though it should be noted that some of these benefits are already occurring and are, therefore, *actual* rather than potential, whereas 'real hazards have not yet appeared'.[57]

Despite the suspicions of Western 'control' and 'exploitation' discussed earlier, it seems undeniably true that crop biotechnology has at least the potential to bring substantial benefits to the poor and the hungry, and the moral significance of that potential cannot be ignored; though there is a danger of over-simplification here, as some would argue that hunger is caused as much by poverty and other political considerations as by direct food shortages. Of course any technology can be misused and its potential benefits dissipated (inorganic chemistry helped Hitler to perpetrate the Holocaust), but the genetic engineering of crops appears to have a moral head-start in offering the prospect of at least helping to meet what many see as the most urgent moral demand of this generation – the prevention of starvation. As one commentator puts it: 'It does seem ironic to invoke possible damage to the Third World as an argument against a technology which can provide the improved crop and animal yields for which the Third World is crying out'.[58]

Furthermore it may be presumptuous and insensitive for those of us in less urgent need of the potential benefits to place too much moral weight upon the possible costs. Less affluent countries and individuals are less able to afford the luxury of such cautious speculation, and to them the moral priorities may appear more obvious. As was noted in Chapter 5, it can be argued that to constrain production in a world of hunger is wrong, even immoral. Failing to proceed with promising technologies, therefore, may be seen to be as morally reprehensible as running the risk of producing undesirable consequences. Harm *may* result from developing these techniques, but harm may equally well result from *failing* to develop them.

There can therefore be no simple straightforward way of weighing potential costs and benefits to produce a clear ethical decision. Priorities have to be assigned to the competing costs and benefits, and those priorities will in turn reflect more fundamental value judgments. This process is well illustrated in the following example.

Gary Comstock, whose objections to BST we discussed and largely supported in Chapter 5, is an American moral philosopher with the rare qualification of having 'walked the beans' (i.e. weeded soybeans by hand) as a child on his grandparents' farm. He concludes another detailed survey of the various costs and benefits of genetically engineered herbicide resistance (GEHR) by adopting a position of 'qualified opposition' to GEHR technology because: '(It) does not seem likely to help us to develop an agriculture consonant with the moral principles I have espoused'.[59] These principles he summarises as follows:

> I have in mind what might be called a traditional ideal of
> what constitutes good farming, an ideal dependent upon
> the agricultural communities where it is still practised. It is
> practised in places like Iowa, where my uncle struggles not

only with weeds and loan payments, but with his own feelings of respect for the soil, and with his father's and mother's and grandfather's and grandmother's ideals of what a farm should be.[60]

This then is the fundamental value judgement by which Comstock assigns priorities to the possible costs and benefits of this technology. He openly admits that this value judgment is 'historically conditioned and qualified' and that 'particular ideals of farming should not automatically bind everyone'.[60] This is a welcome admission, for it is at least debatable whether Comstock's nostalgic, Iowan ideals justify the overriding ethical priority that he assigns to them; they are unlikely to be shared, for example, by victims of famine and drought in the developing world. Value judgments cannot be proved to be right or wrong, but in trying to weigh the costs and benefits here it seems reasonable to attempt to go beyond the boundaries of our 'historical conditioning' and assign a high ethical priority to the undoubted potential which this technology offers for alleviating human suffering.

Conclusions

The genetic engineering of plants raises a number of possible concerns that mainly relate to the extrinsic consequences of this technology in terms of risk and of undesirable socio-economic consequences.

The particular example of the tastier tomato is the least suspect in these respects, as the hypothetical risks associated with it seem negligible and it is unlikely to create much, if any, socio-economic upheaval. While some critics might argue that it is misguided to devote so much research and technological effort to a trivial product for which there is little social need, our view is that this work, while not being particularly significant in this instance, has much potential value because it can be used with a number of other crops of greater social and economic importance.

With herbicide-resistant and pest-resistant crops, the stakes are a lot higher and firm conclusions cannot confidently be reached at this stage. Scientific opinion is clearly divided on the possible risks, and this suggests that a cautious, well-regulated, step-by-step approach is the most responsible way forward. The socio-economic effects are even more difficult to predict, and there is a clear ethical obligation upon those countries and institutions that will profit from the new technology to minimise the losses of the most vulnerable economies and individuals. Overall, however, the likely costs and benefits have to be weighed in terms of ethical priorities and more fundamental value judgments. This process cannot yield a conclusive right-or-wrong answer, but given the present state of knowledge, our view is that work on herbicide-resistant and pest-resistant crops should be further developed in view of its potential benefits for human welfare.[61]

7

The genetic engineering of animals

For the production of human proteins in milk, the obvious choice for transgenic bioreactors – living organisms modified to produce a particular chemical – are dairy animals.

Whitelaw (1995)[1]

Did you hear about the genetic engineer working for a large multinational company who tried to cross a hyena with an oxo cube and ended up making a laughing stock of himself. . .?

Kinsman (1991)[2]

Introduction

As we have seen in Chapters 5 and 6, the genetic engineering of both microorganisms and plants raises a number of ethical considerations. For most people, though, the genetic engineering of animals both brings these ethical issues into sharper relief and poses new questions. In this chapter, we shall continue with our practice of focusing on a few case studies. This will allow us to examine the issues in some depth. The case studies have been chosen to raise a variety of ethical questions.

A final preliminary point: throughout this chapter we shall adopt the convention of using 'animal' to mean 'non-human animal'. The longer phrase can become tiresome. However, using 'animals' to exclude humans is dangerous. It can lead to an assumption that the divisions between non-human animals and humans are greater than is the case. We shall return to this point later in the chapter.

Tracy: the most valuable sheep in the world

Emphysema is a condition in which a person's lungs degenerate. The walls of the tiny air sacs (alveoli), across which oxygen and carbon dioxide are exchanged, break down. The disease is incurable: lungs that have become damaged by emphysema cannot, at least at present, be treated in such a way that they regain their normal functioning. The most common cause of emphysema is cigarette smoking. If a person suffering from emphysema as a result of cigarette smoking stops smoking, their lungs don't deteriorate any further. Continued smoking, however, can lead to a progressive deterioration in lung functioning resulting, eventually, in death.

Emphysema is caused not only by cigarette smoking. It can also result from exposure to other airborne irritants and can be a consequence of asthma and chronic bronchitis. In addition, one person in every few thousand is unfortunate enough to develop emphysema as a result of genetic mutations in their DNA. These genetic mutations lead to a deficiency in the production of a particular protein called α_1-antitrypsin (AAT). AAT protects lung tissue from attack by powerful enzymes that are released by scavenging white blood cells as they go about their normal business of defending the body against attack by invading microorganisms.[3]

Someone suffering from emphysema can only satisfactorily be treated by being given approximately 200 grams a year of AAT. At present this is only available – and then at prohibitive expense – by isolating the protein from extremely large amounts of human blood. However, work by a team of researchers at PPL Therapeutics in Edinburgh is changing this.[4] In 1990, this team succeeded in transferring copies of the human gene that makes AAT into five sheep embryos. Four of the sheep were females and the star performer, called Tracy, produces approximately 35 grams of AAT in each litre of her milk (Figure 7.1).

One might ask why sheep have to be used. Why can't AAT be made by bacteria or other microorganisms in the same way that human insulin is, as we discussed in Chapter 5? The short answer is that it probably could be, but it would be far more difficult. The reason for this is that many human proteins, including AAT but

Figure 7.1. Tracy – the most valuable sheep in the world. Tracy produces large amounts of the human protein AAT in her milk. This protein can help people suffering from emphysema.

excluding human insulin, carry sugar molecules on them (they are said to be 'glycosylated'). In the human body these sugars play a variety of important roles, yet proteins made by bacteria lack them. However, when the AAT gene is inserted into a mammal, such as a sheep, the protein that the mammal makes does have these sugars on it. Adding these sugars to AAT produced by bacteria is not yet feasible. Even if it becomes feasible it is likely to be complicated and expensive.

A second reason for using female mammals, such as sheep, is that the genetic engineers can insert, along with the AAT gene, a control gene that ensures that AAT is made only by the cells in the sheep's mammary glands, which produce milk. This means that the cells in the rest of the sheep do not produce AAT. For the sheep, this is probably a good thing in the sense that it doesn't have a human protein in its blood or nervous system or indeed anywhere except in its milk. From the genetic engineer's point of view being able to restrict production of AAT to milk is ideal. Milk is the perfect way to collect proteins made by an animal. You don't have to harm the animal in any way or even inject it. You simply milk it and then, using straightforward purification techniques, separate out the AAT from the other constituents of the milk.

By 1995, laboratory studies had shown that genetically engineered AAT produced by Tracy and her relatives appears very similar to natural AAT produced by humans. Clinical trials should follow shortly.

Not only sheep have been genetically engineered to produce human gene products in their milk. Mice, rabbits, goats and cattle have similarly been engineered in research projects to produce a variety of proteins. For several years, a near-farcical debate has ensued in the European courts about the reproductive opportunities that should be offered to Herman. Herman is one of the world's first transgenic (genetically engineered) bulls. He is owned by Gene Pharming, a Dutch company that just happened to have a research director called Herman de Boer.

In 1989 Gene Pharming was given permission to try to create, by genetic engineering, a cow that carried the modified version of the human gene for lactoferrin.[5] Lactoferrin is a protein secreted in breast milk which helps to protect babies against infection. The hope was that in its modified form it could help prevent mastitis in cows. The first problem arose when the best efforts of Gene Pharming failed to produce any genetically engineered cows but came up with Herman instead. The second problem arose in 1992 when the Dutch parliament decided to forbid the genetic engineering of large animals except in cases where human lives may be saved and where no alternatives exist. However, Gene Pharming managed to persuade the courts that its genetically engineered lactoferrin would be useful for treating gut infections that often occur in people with cancer. Herman was at last given permission to reproduce, though it took a vote in the Dutch parliament to decide his fate.

Transgenic mice and rats: models for human diseases

Before the PPL team genetically engineered Tracy and her sisters to produce the protein AAT, the Roslin Institute, which works with PPL, genetically engineered mice to produce it. The reason was not that they ever expected genetically engineered AAT to be obtained from mice on a commercial scale – mice don't produce a great deal of milk and milking them is not easy – but rather that mice are excellent experimental animals on which to try such a procedure. In particular, they have a very short generation time and large numbers can be kept easily, cheaply and conveniently in a laboratory.

These advantages, and the fact that a tremendous amount is known about mouse genetics, mean that genetically engineered mice and rats are being used in increasing numbers as models for human diseases. The basic procedure goes as follows. First, genetically engineer your mice (or rats) so that they mimic a human disease. Secondly, study these altered animals either to investigate the disease or to see if it can be alleviated. The hope, of course, is that what is learned about the disease from the genetically engin-eered mice or rats will be applicable to humans.

Examples of human diseases for which genetically engineered mouse models exist include the following:[6-8]

- cystic fibrosis
- severe combined immunodeficiency
- muscular dystrophy
- albinism
- cancers
- Lesch–Nyhan syndrome
- sickle-cell anaemia
- β-thalassaemia
- atherosclerosis
- high blood pressure
- Alzheimer's disease.

The first of these genetically engineered mice was the so-called Harvard oncomouse developed by Philip Leder and his colleagues

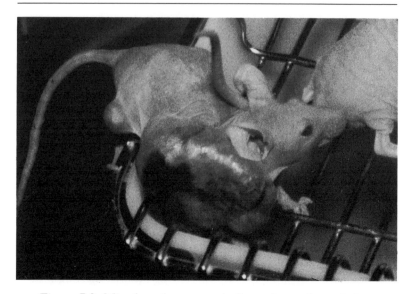

Figure 7.2. Mice have been genetically engineered to mimic a number of human diseases, including various cancers. This photograph shows a mouse in which a cancerous tumour has been induced, though not as a result of genetic engineering.

at Harvard Medical School and patented in 1988. The Harvard oncomouse contains certain human genes, which means that the majority of the individual mice in the strain develop cancers (Figure 7.2). Quite apart from whether or not it is ethically right to genetically engineer mice so that most of the individuals develop cancer, a question to which we shall return, it has to be admitted that the Harvard oncomouse has not been a tremendous commercial success.[7]

Du Pont, the company that funded the research, had invested millions of dollars in the project by 1993. However, by then it had failed to persuade a single pharmaceutical company to sign up for deals involving the mice. Du Pont had hoped that it would be able to charge a royalty on anti-cancer drugs developed through studies using the mice.

However, there are instances where genetically engineered mice are proving more useful. Consider the development of the SCID-hu mouse by the American company SyStemix. A SCID mouse is one that has had its entire immune system knocked out. It can be used to model the human disease severe combined immuno-deficiency, in which a baby is born without an immune system. While he was working at Stanford University with people who had AIDS, Mike McCune, now SyStemix's Vice-President of New Enterprise Research, came up with the idea of developing a hybrid between the SCID mouse and the human immune system.[8] He took embryonic human immune cells, obtained from an aborted fetus, and transplanted them into a SCID mouse. The resulting SCID-hu mouse lacks the mouse immune system but has a functional human immune system.

Because the SCID-hu mouse has a human immune system, it is susceptible to, for example, HIV (human immunodeficiency virus). As a result, the mouse can be used to test possible anti-HIV drugs. SyStemix was able to determine, in just four weeks, the optimal dose of the anti-HIV/AIDS drug AZT, something that it took Burroughs Wellcome four years of clinical trials on humans to establish.

Another instance where genetically engineered mice have proved to be of value is in the development of new therapies for cystic fibrosis. In 1993, scientists from the Imperial Cancer Research Fund in Oxford and the Wellcome Trust at the University of Cambridge found that mice that had been genetically engineered to show symptoms of cystic fibrosis could themselves have their symptoms alleviated through genetic engineering.[9]

Controversy exists, though, as to precisely how valuable or necessary the use of any animals, let alone genetically engineered ones, in medical research is. Some people, including the majority of medical researchers, maintain that their use is essential; others that improvements in alternative approaches (including cell culture, tissue culture and computer modelling) mean that animals are no longer needed for such work.

Pigs for human transplants

Every year many tens of thousands of people around the world are judged to need an organ transplant. Fewer than half of them ever get one, simply because there aren't enough donor organs available. In the UK, 25–30% of the patients waiting for heart or lung transplants die before suitable organs become available to them. Ironically, the problem has got worse in many countries in recent years as driving has become safer. With the more widespread use of seat belts and motorcycle crash helmets, fewer organs are available from accident victims. What is the solution to this problem of insufficient donor organs? One approach currently being investigated is to use genetic engineering on pigs so that their organs can be given to humans.

This research is being pioneered by John Wallwork and David White at Imutran Laboratories in the University of Cambridge, by Jeffrey Platte at Duke University, by a Yale firm called Alexion Pharmaceuticals and by DNX in Princeton.[8,10,11] The basic principle is fairly easily described. Since the mid-1970s, it has been known that transplants to humans from our close evolutionary relatives, such as chimpanzees and baboons, can occasionally prove successful. There are, however, major disadvantages with such transplants. First of all, their success rate is very low; secondly, the animals that give the best results – chimpanzees – are themselves rare; thirdly, the use of primates raises, in many people's eyes, particular ethical problems; and fourthly, patients may need to be given such large doses of drugs to prevent the transplanted organ being rejected that they become susceptible to a huge range of opportunistic infections caused by viruses, bacteria and other germs. Attempts to use animals less closely related to us fail because the organs are rejected. Until the advent of genetic engineering, that is. Today's research aims genetically to engineer pigs so that instead of their hearts or other transplanted organs triggering an immune response when they come into contact with human blood, they are coated with human proteins. In this way, the theory goes, the pig's organs will be unrecognised as foreign when transplanted into a human, rather as the wolf initially fooled Red Riding Hood by wearing her grandmother's clothes.

Reasons for using pigs in this research are that they are domesticated, easy to breed, have large litters and grow rapidly to the size of a large person (in contrast, a large baboon weighs only about 30 kg). In addition, they are surprisingly similar to humans in some

Figure 7.3. Two of Astrid's descendants. These pigs have been genetically engineered to have their internal organs coated in human proteins, thus making them more suitable for transplanting into humans.

important anatomical and physiological respects. The first pig genetically engineered in this way was born on Christmas Eve 1992, and called Astrid. By 1994, there were literally hundreds of pigs genetically engineered for this research (Figure 7.3). By 1995, experiments in both the UK and the USA in which genetically engineered pigs' hearts were transplanted into non-human primates had produced very encouraging results. It is too soon to be sure when clinical trials will start, but John Wallwork and David White think the work may have a major clinical impact by the year 2000.

Genetically engineered farm animals

Whether or not transplants come to use organs obtained from genetically engineered farm animals, it is extremely likely that the extraction of human proteins from the milk of sheep and other animals will become increasingly common in the years ahead. We have already seen that farm animals such as Tracy and Herman are being genetically engineered to produce human proteins. In a related development, Helen Sang and colleagues at the Agricultural and Food Research Council's Roslin Institute in Edinburgh have developed ways of making genetically engineered hens.[12] The idea is that the hens could be made to lay eggs rich in valuable pharmaceuticals, a case, in essence, of hens laying golden eggs. However, aside from the production of pharmaceuticals for human benefit, there are a number of other reasons why scientists are attempting to carry out genetic engineering on animals. Most of these are concerned with ways of boosting the productivity of farm animals.

The largest amount of research has gone into inserting growth genes into fish, pigs, chickens, sheep and cows.[13] The hope is that the altered organisms will grow bigger or faster or produce leaner meat. As is well known, an early example of this work did not turn out as hoped. The 'Beltsville' pigs – named after the US Department of Agriculture research station where they were born – were given a human growth hormone gene. As intended, the animals did grow faster and proved leaner too; a point likely to make their meat more attractive to certain consumers convinced, in these health-conscious days, that animal fats are unhealthy. However, the animals were arthritic, had ulcers, were partially blind and (in the case of the males) were impotent. Sheep that have been genetically engineered with growth hormone genes have proved to be diabetic and suffer high mortality. However, more success has come with fish, where genetically engineered individuals can grow much faster than their natural counterparts.

A different approach to boosting farm animal productivity is being taken by Bernie Wentworth and his colleagues at the University of Wisconsin. They are working to produce genetically engineered turkeys that are unable to produce the hormone prolactin.[14] The reason is that prolactin is the hormone that naturally triggers broody behaviour. Wentworth calculates that turkeys could lay between 15 and 20% more eggs if they did not lapse into periods of broodiness when they want to incubate the eggs they have laid (Figure 7.4).

Figure 7.4.

Finally, it is worth mentioning an attempt by Australian scientists to produce 'self-dipping' sheep. The aim is to genetically engineer sheep so that they produce an insecticide in their skin.[15] Researchers have inserted the gene for an enzyme called chitinase – produced by a number of plants – into mice (which are being used as an experimental half-way house to sheep). Chitinase breaks down the most important chemical (chitin) in insects' skeletons. The hope is that sheep genetically engineered to produce chitinase in their skins will be immune to blowflies, lice and other insect parasites. The main benefit would be a commercial one. At present the Australian sheep industry spends some A\$300 million a year on pesticides. In addition, there is widespread concern at the effects on the health of farmers of the pesticides currently used to dip sheep.

SHOULD WE GENETICALLY ENGINEER ANIMALS?

Having examined some of the uses to which genetically engineered animals can be put, we now turn to ask whether such work is ethically acceptable.

Welfare considerations

Most people would agree that if genetically engineered animals end up suffering significantly then this is unacceptable unless there are, at the very least, extremely pressing reasons for the work. Before looking at particular instances to see if genetic engineering does cause suffering, we need more generally to examine what is meant by animal suffering.

Suffering involves susceptibility to pain and an awareness of being, having been or being about to be in pain.[16] Pain here is used in its widest sense and includes stress, discomfort, distress, anxiety and fear. It is difficult to argue against the contention that vertebrates, and probably certain invertebrates such as octopuses, can experience pain.[17] The extent to which animals are aware of their pain is more open to question. There is little doubt that certain of our closest evolutionary relatives, such as chimpanzees and other apes, have the requisite degree of self-consciousness.[18] Although the extent to which other animals can suffer is contentious, a growing number of biologists and philosophers accept that, at the very least, most mammals, and probably most vertebrates, can. This conclusion is unlikely to surprise anyone who has ever kept a pet or has worked with animals, as a farmer or vet for example.

So does the genetic engineering of animals cause suffering? No overall answer can be given. We need to proceed on a case by case basis. Take, first of all, the case of Tracy (Figure 7.1) and her relatives genetically engineered to produce the protein AAT. There is no doubt that Tracy's physical welfare could not, realistically, be much better. She is worth as much as a good racehorse and treated accordingly. It is true that some might describe her life as a little boring (as is that of any farm sheep) but that hardly constitutes suffering. It is the case that her offspring are taken away from her a week after their birth – on the grounds that her milk is far too valuable for mere lambs to swallow – and some might object to this as causing her mental suffering. However, if there are objections to be made to the genetic engineering that has led to Tracy and others like her, they cannot be based primarily on welfare grounds.

The case of Astrid and the other pigs genetically engineered to be used one day as transplant material (Figure 7.3) is slightly less clear cut. Anti-vivisectionists have claimed that tampering with a pig's immune system could lead to immune disorders or a loss of resistance to infections. Empirically, though, there is no evidence of this to date. Indeed, the success of the research depends, to a large extent, on the pigs being in prime health. Should this research ever take off, the pigs are likely to be killed long before the end of their natural lives, but then most pigs used in the meat industry are killed around six months of age.

There are, though, cases where genetic engineering has undoubtedly led to animal suffering. We have already mentioned (p. 174) the USDA (United States Department of Agriculture) Beltsville pigs, which turned out arthritic, partially blind and with ulcers. One way of viewing this is to believe that, as argued by Russ Hoyle, a contributing editor of the influential magazine *Bio/Technology*: 'To most microbiologists and geneticists, USDA's pathetic pigs, as a research effort of the early 1980s that was quickly discontinued, are simply a failed experiment.'[19] Aside from what some would regard as a certain callousness here, it may be overly optimistic to imagine that suffering resulting to farm animals from genetic engineering is a thing of the past. Genetic engineering of animals is still usually carried out by a rather crude method in which several hundred copies of the gene in question are simply inserted into a fertilised egg. Despite this massive injection, the overall success of the method is rarely more than 2%, meaning that in 98% of cases copies of the injected gene fail to insert themselves into the host's DNA. When the technique is successful there is, as yet, virtually no control over where the genes, or how many of them, insert themselves. In the UK, the Banner Committee, set up to consider the ethical implications of emerging technologies in the breeding of farm animals, concluded in 1995, with regard to the welfare implications of genetic engineering, that:

> Whether a proposed modification will be harmful is presently difficult to predict. The current lack of knowledge . . . contributes to the difficulty and means that any modification risks producing an animal whose welfare is in some way harmed. There are some 50,000 to 100,000 expressed genes

in mammals and the effect of an inserted gene will depend on
its position and on its interaction with other genes.[20]

In the cases we have considered in this section so far, any adverse
consequences of genetic engineering for the welfare of the animals
involved are both incidental and unintentional. In the case, however,
of mice and other animals genetically engineered to act as models of
human diseases, this is not the case. Here the express intention is to
breed animals that develop cancers and other diseases.

In some cases, this may still not result in animal suffering.
Someone may find it distasteful that mice are bred to develop
Alzheimer's disease from an early age. However, this does not
necessarily mean that the mice suffer. Alzheimer's is a distressing
disease, but even in humans it need not cause physical suffering.
Any suffering caused is mental rather than physical, and may be
more significant for the patients' relatives and friends than for the
patients themselves.

However, this line of argument, which would anyway be rejected
by some, breaks down when we consider cases such as oncomice
(Figure 7.2). Common sense, the scientific literature and the
courts have all concluded that these animals suffer as a result of
genetic engineering.[21] Oncomice develop tumours in a variety of
places including mammary tissue, blood, skeletal muscle, the
lungs, the neck and the groin. Tumours can lead to severe weight
loss (40% body weight or more) while large tumours may ulcerate.
In some instances, oncomice have suffered limb deformities as a
side effect of the genetic manipulation.

The specific ethical question raised here by genetic engineering
is whether it is somehow less justifiable to breed mice genetically
programmed to succumb to a particular disease than it is to *induce*
that disease in the mice by other more conventional means. From
the point of view of animal suffering, there is little or no difference
between the two approaches, so is there anything *distinctively*
unethical about the genetic approach? The use of animals for any
medical research clearly raises difficult ethical questions, but it is
important to distinguish these *general* ethical issues of animal
welfare from any specific concerns about the use of genetic
engineering.

Despite this distinction, it is hardly surprising that the development and patenting of oncomice has been attacked by a large number of animal welfare and animal rights movements around the world. In addition, religious organisations are increasingly speaking out against the suffering that humans cause to animals. In some religions, such as Christianity and Islam, the points of view that centre on the animals have only really been put with any strength in recent decades.[22] A number of other religions, though, have a much longer history of according priority to the non-suffering of animals. In Jainism, the concern for *ahisma* (non-injury) goes hand-in-hand with an insistence on a vegetarian diet, while lay members are encouraged to engage only in occupations that minimise the loss of life. Within Jainism, it is the monastic practice to carry a small broom with which gently to remove any living creature before one sits or lies down.[23] In Buddhism too, there has traditionally been a strong emphasis on animal well-being.

Having examined the extent to which the genetic engineering of animals may affect their welfare, we now need to turn to an even more contentious issue, namely whether animals have rights.

Do animals have rights?

It is often held by philosophers that the notion of 'rights', whether or not applied to animals, is problematic. As Mary Midgley notes: 'This is the really desperate word. As any bibliography of political theory will show, it was in deep trouble long before animals were added to its worries'.[24] The reason for this apparently rather despairing view is partly that the word is used in a variety of overlapping ways, partly that historically it has been inextricably linked with the notion of *legal* rights and partly that rights are commonly taken to be correlative with duties while many philosophers follow Kant in rejecting the notion of duties to animals.

It is, of course, possible to argue that animals 'have rights', though this may not be the clearest or most helpful way of highlighting the ethical issues concerning our treatment of animals.[25] A different approach is to by-pass the question and focus instead on the extent to which our use of animals is often 'species-ist'. It was the Australian philosopher Peter Singer who, in 1975, produced the first really sustained argument that most humans are guilty of 'speciesism'.[26] The term is, of course, chosen to have echoes of 'racism' and 'sexism' and so to refer to attitudes and behaviours that once were commonly deemed acceptable but are now generally considered morally unsustainable. Put at its most succinct, it is of little significance, the argument goes, that humans belong to a different biological species from, say, chimpanzees, dogs, farm animals and laboratory mice; we do not have the right to treat such species merely as we choose and for our own ends.

Think of the conditions we normally require before humans are permitted to be used as research subjects.[27] We require that two conditions be met: first, that the participating individual gives their informed consent; secondly, that there is no intent to do harm to that individual. The second of these conditions is inviolate. The first can only be overturned when patients are unable to give their consent, for instance because they are babies or in a coma, when it can be given on their behalf. Further, most of us are not, generally, persuaded by the utilitarian argument that these conditions can be overturned if a number of other people would benefit. This, of course, is why most people hold that, in everyday language, humans have rights. For example, nowadays few of us would be persuaded that subjecting even a few people to slavery is acceptable whatever the beneficial consequences that might result for the rest of us. In the language of Chapter 3, there are intrinsic objections to such things as slavery and experimenting on people who object to this, whatever the benefits.

The problem is, why do most people hold that it is not permiss-ible to subjugate people into slavery or to experiment on them without their consent when we regularly do these things to non-humans, including even our closest evolutionary relatives, namely chimpanzees and other mammals? We can also note that if one adopts solely the criterion of suffering to decide whether or not an organism should be used for human ends, not only would the use of most animals for medical research cease, but a case could be made, abhorrent as it sounds, for mentally handicapped new-born human infants to be used for such research, on the grounds that such infants are arguably incapable of suffering yet are physiologically closer to self-conscious sentient people than are the laboratory animals presently used.

In the otherwise excellent 1991 Report of the UK Working Party of the Institute of Medical Ethics into the ethics of using animals in biomedical research, it is argued that this conclusion can be rejected on the grounds that: 'being of the human species may be a *sufficient* condition of being awarded enhanced moral status . . . . Possessing the *nature of a rational self-conscious creature* may be sufficient for being awarded this status even though this nature may be impaired or underdeveloped in the individual case'.[28] To many people this argument comes across as special pleading. Why is being of the human species a sufficient condition of being awarded enhanced moral status? Some people would argue, from a religious perspective, that there is something qualitatively different about being of the human species (a condition held by some to begin at the moment of fertilisation). We have already discussed, in Chapter 4, how some religions see humans, alone of the created order, as being made 'in the image of God'. However, as we also discussed in Chapter 4, there have been significant shifts in recent decades within even such traditionally anthropocentric religions as Judaism, Christianity and Islam with regard to how humans should treat the rest of the created order. For example, many people with a religious faith would now be unhappy with the following argument proposed by an Anglican priest in 1992:

> Suppose a life has to be sacrificed through shortage of
> facilities: either a healthy chimpanzee, or an individual with
> Tay–Sachs disease who will inevitably die with a few months.
> 'Moral individualism' asks which life is of more value *to the
> individual concerned* (infant or chimpanzee), and opts for the
> latter; Christian faith asks which life is of more value *to God*,
> and its answer is understandably different.[29]

It is evident that there is still considerable controversy, whether or not one has a religious faith, both over whether animals have rights and whether animals are qualitatively different from humans in some way that allows us to treat them differently. Before attempting to see whether any guidelines can be suggested as to when it is ethically acceptable to carry out genetic engineering on animals, we need to turn to the question of how genetic engineering may alter the essence of an animal.

How may genetic engineering change the essence of an animal?

So far we have mainly considered whether or not it is acceptable for humans to cause suffering to animals. However, it is not the case that the only objection that can be raised against the genetic engineering of animals is that it may cause suffering. Consider the example we described earlier where genetic engineering is being used to prevent broodiness in female turkeys, potentially boosting productivity by around 15–20%.[14] Let us assume that the genetic engineering works and that the female turkeys do not suffer in any way, is this ethically permissible?

Now this work seems, in some sense, to be altering the 'essence' or 'being' of a female turkey. Part of being a female turkey is, presumably, to have the capacity to brood one's eggs. Is it acceptable to take away this ability, even if no suffering results? It will come as a surprise to few readers to know that Jeremy Rifkin, Director of the Foundation on Economic Trends and a tireless protagonist against genetic engineering, argues that such genetic engineering is unacceptable: 'I will make this an issue . . . This is a serious violation of the intrinsic value of the creature. It's ethically unconscionable'.[30]

However, we should recognise that the conventional breeding of turkeys has already led to a situation in which the birds are unable to mate. Instead, the males have to be 'milked' by hand – to obtain their semen – and the females artificially inseminated. That this is the case does not, of course, mean that it is ethically acceptable to genetically engineer turkeys so that the females are unable to brood. Rather it reminds us that conventional breeding can also change the essence of an animal. Indeed, the entire history of the domestication of farm animals is, in a sense, a history of the accumulation of successive small changes to their essence. Over the millennia, farm animals have become more tractable and more tolerant of overcrowding. Further, at present, farmers try to 'shock' female turkeys out of broody behaviour by exposing them to bright lights or by making them stand on wires, so that they are unable to settle down and brood.[14] It could be argued that genetically engineering turkeys so that the females do not show broody behaviour will be to the benefit of the animals' welfare. Of course, this line of reasoning is open to disputation on the grounds that two wrongs don't make a right. Suppose someone argued that a good way to reduce the high incidence of sexually transmitted diseases among child prostitutes would be to issue them all with free condoms (analogous to improving the welfare of domestic turkeys by eliminating brooding behaviour through genetic engineering). Most of us would reject the argument, holding that the exploitation of children for prostitution is unacceptable and should be prevented (analogous to improving standards of conventional farming so that female turkeys can make proper nests and brood at least some of their eggs).

The turkey example is illustrative of the more general question: should we genetically engineer animals to reduce their natural capacities? For example, should we reduce the capacity of animals to feel pain?[31] At the other end of the scale, should we attempt to improve animal welfare by increasing the natural capacities of animals? For example, should we enhance disease resistance by genetic engineering?[32]

The philosopher Alan Holland has argued that: 'We ought to be very concerned by any developments which would diminish the general level of freedom of sentient animals'.[33] Holland argues that diminishing the general level of freedom of sentient animals may be a secular equivalent of blasphemy. A related point is that in Kantian terms, as we outlined in Chapter 3, it may be acceptable for us to use an animal's ends as our own (e.g. using a sheep to produce wool and lambs), but it is unacceptable for us entirely to ignore an animal's ends and instead use it solely as an instrument by which we attain our ends.

We can put some flesh on the bones of this argument by thinking in terms of the *telos* of an organism. The *telos* of an organism derives from the Aristotelian use of the term and can roughly be understood as the fulfilled state (or end or goal) of the organism. A number of writers, in addition to Holland, have argued that it is wrong for us to violate the genetic integrity or *telos* of organisms. However, a major problem here is that *telos* is a rather imprecise concept, used in different ways by different writers. A pig's *telos*, for example, can be thought of as its essence, beingness, inherent nature, purpose, ends, needs, interests, or even (most comprehensively) as its 'pigness'. This elastic meaning makes it virtually impossible to decide whether domesticated animals at least have a *telos*, and if they have whether it is wrong or even possible to try to modify it. A more useful approach may be to concentrate on the distinction between instrumental and intrinsic value.[34]

In essence, something is of intrinsic value when it has a worth in itself, irrespective of its usefulness for humans. To illustrate the instrumental attitude, consider the arguments of Jan Heideman who carried out the pioneering work with Amy Moser that led to the development of genetically engineered mice. According to Heideman, three of the advantages genetically engineered mice offered over genetically engineered rats were that 'a rat's skin cannot be peeled off as can the skin of a mouse', 'a rat does not fit on the stage of a standard dissecting microscope' and 'the transport of rats can become a real logistical problem'.[35] For most people, such an attitude is sufficient to cause them to question the adequacy of the instrumental view.

The fullest attempt to work out what taking seriously the intrinsic worth of non-humans would mean is presented by Paul Taylor in his book *Respect for Nature: A Theory of Environmental Ethics*.[36] His argument is likely to be attractive to the many who argue against a purely instrumental approach to nature. Taylor develops a bio-centric (life-centred) as opposed to an anthropocentric (human-centred) environmental ethics. He argues that all living organisms possess inherent worth, and he explores the consequences of this for the resolution of conflicts between humans and other species when there is competition for limited resources.

With specific reference to the genetic engineering of farm animals, the UK Banner Committee produced in 1995 a set of three principles that it maintained should be followed before any new technology, not just genetic engineering, is used in the breeding of farm animals:

(a) Harms of a certain degree and kind ought under no circumstances to be inflicted on an animal.

(b) Any harm to an animal, even if not absolutely impermissible, nonetheless requires justification and must be outweighed by the good which is realistically sought in so treating it.

(c) Any harm which is justified by the second principle ought, however, to be minimized as far as is reasonably possible.[37]

The main problem, of course, in using the first of these principles is to agree on the particular harms that ought not to be inflicted on an animal. The Banner Committee argued that it would be unacceptable, through genetic modification, for a scientist to increase the efficiency of food conversion in pigs by reducing their sentience and responsiveness, thereby decreasing their level of activity, on the grounds that 'the proposed modification is morally objectionable in treating the animals as raw materials upon which our ends and purposes can be imposed regardless of the ends and purposes which are natural to them'.[38] As we have already noted, the main difficulty in accepting the conclusions of this particular example is that several thousands of years of selective breeding in pigs have already led to an increase in their efficiency of food conversion by reducing their responsiveness, and possibly also their sentience.

Our object in raising this difficulty is not to reject the principles proposed by the Banner Committee but to point out that it may well prove impossible in this area to produce any absolute principles that take account of all eventualities. While this may be disconcerting to some moral purists, anyone who lives in the real, political world is unlikely to be too concerned, for in such a world progress is usually incremental and by consensus rather than by the whole-hearted adoption of clear-cut ethical principles.

Should we eat genetically engineered animals?

We have already argued, in Chapters 5 and 6, for the labelling of genetically engineered foods. Such labelling, we suggested, would be of value to several categories of consumer:

In the beginning the Lord created heavens and the earth and all therein.

But scientists said let the seas be populated by fish containing human, mouse and rat genes.

And Hoechst and Monsanto said let the land produce plants and trees bearing fruit containing bacteria and virus genes.

And Ciba Geigy said let corn grow containing scorpion genes.

And Amoco said let there be tobacco containing hamster genes.

And Zeneca made all kinds of fruit and vegetables and they did not rot.

By the seventh day the work of the Lord had been undone and the companies saw they had recreated life for their own ends.

And the Parliament of Angels blessed the seventh day and made all life patentable.

Figure 7.5. Before the 1995 vote in the European Parliament on the patenting of genes, Greenpeace sent a postcard with this message to all the Members of the European Parliament.

- Those who object to genetic engineering of any sort
- Vegetarians who might not wish to eat plants containing animal genes
- People with allergies to particular gene products
- Those who prefer certain genetically engineered foods (e.g. those who prefer cheeses made with genetically engineered rather than animal rennet).

The need for such labelling is becoming increasingly clear as the number of genetically engineered foods increases (see Figure 7.5). Indeed, in 1992 Jeremy Rifkin launched his Pure Food Campaign with its accompanying logo (Figure 7.6).

We shall now consider three further questions in connection with eating foods resulting from the genetic engineering of farm animals:

- What should be our attitude to farm animals into which an unsuccessful attempt has been made to insert a human gene?

Figure 7.6. The logo of Jeremy Rifkin's Pure Food Campaign.

- What should be our attitude to farm animals into which animal genes from other animal species have been inserted?
- What should be our attitude to farm animals into which human genes have been inserted?

These questions were addressed by the 1993 UK Polkinghorne Committee.[39] This was set up because in 1990 the UK Advisory Committee on Novel Foods and Processes received a submission from PPL (the firm that produced Tracy) concerning certain sheep that fell into the first of the three categories outlined above. These sheep were ones where attempts had been made to genetically engineer them to carry the human gene responsible for the production of factor IX – a protein involved in blood clotting and needed by many haemophiliacs – but where the procedure had failed. In other words, these sheep did not contain the desired gene. This situation comes about because, as we saw earlier (p. 177), only a small percentage of attempts to introduce foreign DNA into mammals succeeds. However, whether or not the procedure has been successful cannot be ascertained at the time it is carried out, only subsequently. So attempts to produce a dozen or so genetically engineered sheep, for example, typically end up with many hundreds of perfectly normal sheep.

It is hardly surprising that the Polkinghorne Committee recommended that such sheep should indeed be allowed to enter the food chain. However, it further recommended that no labelling should be required. Subsequently, though, the UK Food Advisory Committee took the view that, if used for food, the food should be clearly labelled. PPL has denied that it ever intended to put such animals into the human food chain. Politically such a decision is understandable, though some might regret the loss of perfectly good meat for human consumption, identical in every respect to normal lamb and mutton and containing no genes of human origin. This example further highlights the problems of labelling, for how exactly could the meat have been described? 'Product of unsuccessful genetic research' or a similar wording would be unlikely to attract queues of eager customers.

We can now consider farm animals into which other animal genes have been inserted. The major concern here comes from certain religious groups. For example, the Union of Muslim Organisations of U.K. & Eire maintains that: 'From an Islamic point of view, flesh of "transgenic" animals, flesh of animals fed with genes from other animals whose flesh is forbidden, flesh of animals fed with a meat diet, fruits and vegetables produced by the introduction of animal genes are all forbidden to Muslims'.[40] A Sikh view is that: '. . . a Sikh is forbidden according to religion to eat cow's meat. Hence if there is a transfer of cattle genes to other animals e.g. sheep & pig, it could lead to objections'.[41]

Finally we can turn to farm animals into which human genes have been inserted and to possible concerns about cannibalism. At one pole are those who argue that eating an animal, or a plant, into which a human gene has been inserted has nothing whatever to do with cannibalism. Cannibalism is about eating human flesh, not eating minute amounts of DNA that once came from just one of the 80000 or so human genes and is now merely a copy of that original human gene. In any case, the human genes that are likely to be inserted into animals are almost bound to be more than 90% the same as animal genes there already – we share a very high proportion of our DNA with other mammals. Further, every baby who breast feeds eats large amounts of another human's (i.e. its mother's) DNA.

Those who object to inserting human genes into animals that are subsequently used for human consumption may argue that the parallels with cannibalism cannot so lightly be dismissed. After all, just because a baby less than a year or so old does certain things with its mother doesn't make it right for the rest of us to do those same things with its mother. Further, the argument, expounded in some detail in Annex G of the Polkinghorne Committee Report, that there is virtually no chance of our eating the original human gene, but merely a copy of it, can be criticised on two grounds. First, in some people's eyes, a minute probability, even one as vanishingly small as the 1 in 10^{55} suggested in Annex G, is still non-zero. Secondly, does the fact that only copies of human genes are eaten make it any better? A copy of a human gene is chemically identical to it. If you, the reader, attempt to publish a copy of this book under your name, on the grounds that this is somehow significantly different from the original text we wrote, Cambridge University Press, the two of us and the courts are unlikely to be impressed with your argument.

Perhaps inevitably, given such a diversity of views on the acceptability of allowing human genes to be incorporated into animals that are subsequently eaten, the Polkinghorne Committee fell back on labelling as the solution: 'We recognise that many groups or individuals within the population object on ethical grounds to the consumption of organisms containing copy genes of human origin. We therefore *recommend* that food products containing such organisms should be labelled accordingly to allow consumers to exercise choice'.[42]

Our own view is that, although labelling is not always as unproblematic as often assumed (see pp. 127–9), it may indeed offer the best way forward in such circumstances. Even so, labelling tends to by-pass rather than directly tackle the ethical issues: hanging a label clearly indicating 'SLAVE' around the necks of certain people in a slave market certainly offers 'consumers' information and free choice but does not address the moral question of slavery.

It should, further, be realised that, even given genuine and detailed labelling, the transfer of human and animal genes into animals and plants that are subsequently eaten will almost certainly mean that, sooner or later, most of us, unless we exercise the strictest control over what we eat, will end up eating these genes. The reason for this is that the way in which today's food industry operates is such that, in a very real sense, none of us can be absolutely certain what we are eating if we buy, for example, a hamburger or a packaged soup. A rather depressing analogy is afforded by oil pollution. It gets everywhere. There is nowhere on the Earth's surface, even Antarctica or the top of the Himalayas or the middle of the Pacific Ocean, that is entirely unaffected by oil pollution.

Do we need genetically engineered animals?

The final question we need to consider in this chapter is 'do we need genetically engineered animals'? We shall begin with Tracy.

As we have seen, it is possible that the protein AAT, which Tracy and her relatives secrete in their milk, could one day be made by microorganisms, such as bacteria, though the procedure would almost certainly cost more and might be less effective. It is perfectly possible, given a free market, that two forms of genetically engineered AAT may eventually exist – one made by sheep, the other by microorganisms. Which is commercially more successful would then depend on a combination of individual consumer choice, public attitudes, prices and regulations. For the moment, though, it is difficult in our view to sustain a strong argument against the genetic engineering of farm animals to produce proteins that can be produced with little or no cost to the animals and very significant benefits to certain people.

Turning now to the possibility of using genetically engineered pigs for human transplants, the issue is more complex.[43] For a start, better health education might lead to less of a demand for hearts and other organs. Then if only more of us carried donor cards and gave permission for our dead relatives' organs to be used in transplants the need for genetically engineered animals to meet the shortfall would be reduced. Finally, though it is still too early to be sure, significant advances are being made in artificial (metal and plastic) organs. Nevertheless, these arguments against the need for genetically engineered pig organs should be set alongside the fact that thousands of people currently die each year through a shortage of donated organs.

In the case of mice genetically engineered to act as models of human diseases, it is difficult to generalise as to the need for this. Those in favour of such work argue that many of the human diseases that mice have been genetically engineered to exhibit are ones where *in vitro* studies, on their own, are inadequate. Many of the more important questions, it is widely argued by biomedical researchers, can only be addressed *in vivo*.[6] So far, however, the practical gains from much of this work have been extremely limited.[21,44]

Turning finally to the genetic engineering of farm animals to increase productivity, it all comes down to what one understands by the word 'need'. It is possible that there is little that the genetic engineering of farm animals will achieve that could not also be achieved, given time, by conventional breeding. But even if it proves to be the case that genetic engineering alone can significantly enhance the productivity of farm animals, is that necessary? The argument that genetic engineering of farm animals will help feed the extra 90 million people that are alive on the Earth each year is less convincing than the equivalent argument in Chapter 6 for plants, partly because most people in the world eat little meat and partly because animals are ecologically far less efficient as a food source than are plants. The strongest arguments, it seems to us, in favour of genetically engineering farm animals to increase their productivity are where this can be done by genetically engineering resistance to a disease or, as we discussed earlier (p. 175), to attack by a pest, such as blowflies and other insect parasites.

A final point needs to be made before we close this chapter with our conclusions. Suppose I object to a particular instance of the genetic engineering of animals: my objection can take a number of different forms. It would be one thing for me to decline to buy, for example, turkeys genetically engineered to make them less broody; another thing for me to campaign for the labelling of such turkeys to be mandatory; another thing for me to campaign for the sale of such turkeys to be illegal (thus attempting to prevent anyone from buying them); another thing for me – assuming that the sale of such turkeys is permitted by the law – to pursue a course of non-violent protest at this (for instance, by picketing my local supermarket); and yet another thing for me to plant incendiary devices at the homes of people who work in supermarkets that, in accordance with the law, sell such turkeys. In other words, ethical decisions can be taken at a number of levels, just as they are taken by different collections of people – individuals, families, local communities, local governments, individual nation states and international bodies – and the way in which objections are expressed can in turn raise further ethical questions about appropriateness and justification.

Conclusions

In this chapter, we have considered a number of examples of the genetic engineering of animals, examining both the arguments in favour of such work and the concerns that they raise. There are, in addition, other concerns about, as there are other benefits of, the genetic engineering of animals that we have not considered in this chapter. For example, there is a significant risk that the widespread use of genetically engineered fish might lead to the sorts of ecological problems we discussed in Chapter 6 in the context of the widespread use of genetically engineered crops.[45] However, the examples we have examined in this chapter give a good indication of the range of ethical issues that currently arise from the genetic engineering of animals.

In our opinion, the genetic engineering of animals needs to be examined on a case by case basis.[46] The main questions that need to be considered are:

- What, if any, suffering is caused to the animals?
- Are any changes to the nature (or essence) of the animals acceptable?
- How significant are the benefits that may result?
- How likely are benefits to result?
- Could these benefits be achieved without the genetic engineering of animals, and if so how easily and at what, if any, additional cost?

- How would the average person in the street feel about the work if he/she was fully informed about all the issues involved? (This is a stricter question than merely asking how most people feel about it – see our discussion of consensus conferences on p. 242.)
- How would the work affect particular groups of people, especially minorities (e.g. those with cystic fibrosis, vegetarians, Muslims)?
- Are there any significant environmental or safety considerations?
- What are the likely economic and social consequences?

We would have liked to end this chapter with a firm set of conclusions about specific instances of the genetic engineering of animals. At its simplest, this would have consisted of just two lists: one of approved examples, one of disapproved examples. However, uncertainties in the answers to the questions we pose in the previous paragraph mean that we have not been able to be so clear-cut.

Nevertheless, our view is that there are examples of the genetic engineering of animals that fall pretty clearly into one or the other of these lists. The genetic engineering of sheep, such as Tracy, to produce life-saving pharmaceuticals is not only morally acceptable, but also morally incumbent on us to pursue, unless alternative treatments arise. The genetic engineering of female turkeys to reduce broodiness, however, is ethically questionable as it embodies an excessively instrumental view of living creatures. The argument that the genetic engineering would both increase productivity and reduce the frustration, even suffering, of the birds is unconvincing. All poultry should be able to engage in nest building and at least some brooding behaviour. Finally, we are not convinced that the genetic engineering of large numbers of mice to develop painful cancers is acceptable given the extremely limited therapeutic value such animals have so far had.

8
The genetic engineering of humans

What they were envisaging in principle was that in due course it could become possible to extract from normal human cells the sequence of DNA that was missing from or wrongly made in the patient. Once isolated these could be used as the pattern, the template, for the synthesis by bacterial enzymes of numerous replicas of itself. This is acceptable as a possibility in the foreseeable future. The next step would be the crucial and probably impossible one: to incorporate the gene into the genetic mechanism of a suitable virus vehicle in such a fashion that the virus in its turn will transfer the gene it is carrying to cells throughout the body and in the process precisely replace the faulty gene with the right one. I should be willing to state in any company that the chance of doing this will remain infinitely small to the last syllable of recorded time.

Burnet (1973)[1]

Today we can identify some of the genes which, mutated, are responsible for cystic fibrosis, Duchenne muscular dystrophy, adenosine deaminase deficiency and several other hereditary disorders. More will be discovered. Already we are beginning to be able to moderate some of these diseases. Fairly soon we should be able to introduce undamaged genes into the germ cells of afflicted patients, and so correct such disorders for generations to come. . . . In time we shall learn the genetic bases, such as they may be, of social inadequacy, criminality, and other behavioural aberrations – even, perhaps, of aspects of intelligence and creativity.

Postgate (1995)[2]

Introduction

This chapter concerns itself with the genetic engineering of humans, that is, with the intentional change, through much the same techniques as we have discussed in previous chapters, of the genetic material in some or all of a person's cells. This means that we are not concerned primarily with DNA fingerprinting, with genetic screening (unless this takes place as a necessary prelude to genetic engineering) or with issues to do with the patenting of human genes. However, as these three areas arise in connection with genetic engineering, we shall briefly review them before turning to genetic engineering *per se*.

DNA fingerprinting

DNA fingerprinting (also known as DNA profiling) has been around since 1984, the technique having been discovered by Alec Jeffreys of Leicester University.[3] Essentially it relies on the fact that no two people, unless they are identical twins, have the same order of bases in their DNA. This means that for each of us, our individual DNA is unique. Further, only a few cells are sufficient for a sample of our DNA to be obtained. A single human hair or a tiny drop of blood, semen or saliva are sufficient to identify someone. The potential for crime detection is clear. In addition, we each inherit half our DNA from our mother and half from our father. DNA fingerprinting can, therefore, be used to settle paternity disputes and other disagreements to do with genetic relatednesses.

One of the first cases to be solved through DNA fingerprinting concerned the rape and murder of two 15-year-old girls in Leicestershire.[4] A 17-year-old youth confessed to the crime. However, Jeffreys was able to show that the youth was innocent and that both girls had been raped by the same man. The police then formed the hypothesis that the person responsible lived locally, since although the crimes had happened three years apart, they had taken place within 500 metres of each other. Eventually the police carried out DNA fingerprinting on blood samples from almost every man between the ages of 16 and 34 who lived in the three villages nearest the crime. Only two people refused to be tested: one was a Jehovah's Witness (many Jehovah's Witnesses believe that the taking of blood for any reason is wrong); the other was the man who was eventually found to be guilty. However, it took some time to catch him because he persuaded a workmate to go in his place to give the blood sample. It was only when his friend got drunk in a pub and mentioned the story that the deception was realised. The next morning both men were arrested at 5.30 a.m. DNA fingerprinting showed that the man who had avoided giving a blood sample was indeed guilty of both rapes. He is now in prison, having been sentenced to life imprisonment.

As is often the case in science, initial euphoria about the power of a new scientific technique was soon questioned.[5] Doubts about the value of DNA fingerprinting have centred on three main issues. First of all, any laboratory can make – and some have made – mistakes in the labelling of samples. Secondly, initially it wasn't acknowledged that reading a DNA fingerprint is not always totally unambiguous. Errors of interpretation do happen. Thirdly, some of the estimates of the probability of a suspect having the same DNA fingerprint as that identified from a sample of blood were seriously under-calculated. By now DNA fingerprinting is widely recognised as a powerful tool, yet, like any tool, it is fallible.

Separately, ethical concerns have been raised about the use to which DNA fingerprints may be put. A DNA fingerprint is, effectively, an identity card. Civil rights groups have raised questions about whether it is right to have national databases of DNA fingerprints, as is becoming increasingly common in a number of countries. There seems little doubt that in a justly governed country DNA fingerprinting can play a valuable role in such morally acceptable activities as solving crimes and determining genetic relationships so as to decide whether particular individuals should be allowed entry into the country to live with their claimed relatives. Problems are more likely to arise when a corruptly governed country has a high proportion of its inhabitants' DNA fingerprints on file. DNA fingerprints can be rather like the bugs used in 'phone tapping and the videos used in the surveillance of shopping malls or large car parks. They reduce privacy. What would it be like to live in a society with absolutely no privacy? This question has been addressed by the philosopher Jonathan Glover and by numerous science fiction writers.[6]

Genetic screening

Genetic screening presently occurs when a search is made to identify individuals who may be susceptible to a serious genetic disease or who may be at risk of having children with that genetic disease. For example, consider Huntington's disease (chorea). This is a distressing condition for which there is no cure. Males and females are equally likely to suffer from it and the disease usually manifests itself in people aged between 30 and 50. The person finds motor co-ordination difficult so that their hands shake and they have problems with balance. As the condition worsens they find it more and more difficult to look after themselves. Death generally results within five to ten years of the onset of symptoms.

Recently a test has been produced that can tell a person with a high degree of confidence (though not with absolute certainty) whether they have the affected gene that causes Huntington's chorea. If they do, there is approximately a 50% chance for each of their closest relatives (parents, siblings and offspring) that they too have the gene. The existence of this test, which allows genetic screening to take place, throws up a number of ethical dilemmas. For example:

- Should the test be made widely available, given that there is no treatment or cure for Huntington's chorea?
- Should the results of a test be made available to that person's relatives?
- Should the results of a test be made available to that person's insurance company or employer?

The ethical problems and opportunities raised by genetic screening are discussed in detail in the 1993 Nuffield Council on Bioethics publication and elsewhere.[7] At the time of writing, few countries have even begun to address the ethical and legal issues that arise from genetic screening. However, in April 1995 the US government's Equal Employment Opportunities Commission ruled that, under the 1990 Disabilities Act, it was illegal for employers to discriminate against people on the grounds that they have a form of a gene that could lead to disease.[8] This ruling, though, does not apply to insurance companies.

As we have already indicated, genetic screening lies somewhat outside the specific domain of genetic engineering and there isn't space here to get into all the detailed complexities of the ethical issues. However, one way forward may be for people to agree to operate on the principle of the 'veil of ignorance' proposed by John Rawls.[9] Rawls was concerned to develop a principle that would allow one fairly to decide in a society how to distribute resources (goods in short supply) among people. His position is best indicated by an extended quote:

> Somehow we must nullify the effects of specific contingencies which put men at odds and tempt them to exploit social and natural circumstances to their own advantage. Now in order to do this I assume that the parties are situated behind a veil of ignorance. They do not know how the various alternatives will affect their own particular case and they are obliged to evaluate principles solely on the basis of general considerations.
>
> It is assumed then that the parties do not know certain kinds of particular facts. First of all, no one knows his place in society, his class position or social status; nor does he know his fortune in the distribution of natural assets and abilities, his intelligence and strength, and the like.
>
> . . .
>
> Thus there follows the very important consequence that the parties have no basis for bargaining in the usual sense. No one knows his situation in society nor his natural assets, and therefore no one is in a position to tailor principles to his advantage.[10]

In the context of genetic screening, this would mean working out policies with regard to employment, insurance and so on that people would regard as fair without their knowing the results of their own screening.

Should we patent human genes?

The final preliminary question to which we shall turn is the question of whether or not it is right to patent human genes. Obviously questions to do with the ethical implications of patenting are, logically, distinct from questions to do with the ethical implications of genetic engineering. However, the practical reality is that the potential for money to be made from genetic engineering has led to a rush to patent human and other genes. We have already examined the ethical issues that arise from patenting genetically engineered plants (pp. 158–60). The only new emphasis that arises when considering the patenting of human genes is that some people feel that the buck stops here. In other words, they may grudgingly accept patenting of genes from microorganisms or plants – possibly even from animals – but they aren't prepared to accept the patenting of human genes.

The fundamental argument in favour of patenting is that it rewards those who have put a lot of time, effort, ingenuity and money into the invention of a new product or process.[11] For a finite length of time (typically 20 years, though the exact length of time varies in different countries), a patent allows the inventor a monopoly right to exploit the patented invention. After this period, the patent ceases. It has been argued that without the patenting of human genes private companies may be far less inclined to invest the huge amounts of money needed to develop pharmaceuticals and treatments based on the new technologies. For example, in 1994 Myriad Genetics, Utah filed a patent on BRCA1, the human gene responsible for almost half the inherited cases of breast cancer and for most cases of ovarian cancer. Mark Skolnick of Myriad Genetics said: 'If it's not patented, you won't get some group to spend money to develop it, and you won't get a high-quality, inexpensive test'.[12]

However, there are those opposed to patenting.[13] For example, Myriad Genetics had been collaborating in its research on the genetic basis of breast cancer with a British team, headed by Mike Stratton at the Institute of Cancer Research. The British team pulled out of this collaboration in 1994 because of disagreements over the ethics of patenting human DNA. To quote Mike Stratton:

'We do not believe pieces of the human genome are inventions; we feel it is a form of colonisation to patent them . . . I don't think it is appropriate for [such a gene] to be owned by a commercial company because, in contrast to an academic organisation or charity, there is inevitably a demand for profit'.[12]

Those opposed to the patenting of human (and other) genes advance a number of arguments:

- It is wrong to patent life and this means that it is wrong to patent either whole organisms or their genes; the very idea is absurd, obscene or blasphemous
- Patenting reduces the exchange of information among researchers
- Patenting encourages researchers to target their efforts where money is to be made, rather than where work is most needed.

Those in favour of the patenting of human (and other) genes advance the following main arguments:

- Patenting is right; it rewards the investment and ingenuity of those who develop new products
- Without the patenting of human genes there will be fewer benefits to health than would otherwise be the case
- In the absence of patenting, firms would resort to greater secrecy to protect their investment
- Patents do not interfere with pure research since experimental use of an invention does not constitute patent infringement.

Our own view is that despite legal wranglings,[14] the patenting of human genes raises no new ethical issues beyond those already raised by the patenting of other genes, and that the patenting of any gene differs little from the patenting of any product or process. Patents are held for only a limited time and there is a possibility that the patenting of human and other genes might allow some financial benefits to accrue to less developed countries, as these are where the global centres of genetic diversity lie.

Somatic and germ-line therapy

It is helpful to distinguish between two classes of cell found in our bodies:

- *Germ-line cells*: these are the cells found in the ovaries of a female and the testes of a male that give rise, respectively, to eggs and to sperm
- *Somatic cells*: these are all the other cells in the body (for example, the cells of our digestive system, nervous system, cardio-vascular system, lungs and skin are all somatic).

The importance of this distinction is that any genetic changes to somatic cells cannot be passed onto future generations. Changes, on the other hand, to germ-line cells can indeed be passed onto children and to succeeding generations. We shall explore the ethical significance of this distinction later in this chapter, but common sense points out that changes to germ-line cells may have wider implications than changes to somatic cells.

Cystic fibrosis

Cystic fibrosis is the most common genetic disease in a number of countries, including the UK, where approximately 1 in every 2000 babies suffers from it. Someone with cystic fibrosis has severe breathing problems and typically suffers from lung infections. Their digestion is poor, they may develop diabetes and they produce abnormally salty sweat.

It has long been known that cystic fibrosis is an example of an autosomal recessive condition. By autosomal is meant that the gene responsible is on neither the X nor the Y chromosome, but on one of the other, non-sex chromosomes. (Humans have 44 autosomal and two sex chromosomes.) By recessive is meant that an affected person has to have two faulty versions of the gene for the condition to occur. In the great majority of cases, this is simply because both the copy they inherited from their mother and the one from their father are each faulty. Although only 1 in 2000 babies in the UK is born with cystic fibrosis, approximately 1 in 24 adults are hetero-zygous for the condition, meaning that they have one working copy and one faulty copy of the gene. Because the faulty copy is recessive, heterozygotes are perfectly normal and have none of the symptoms of cystic fibrosis. However, if they have children with another person who is heterozygous for the gene, there is a one in four probability that any child they have will have cystic fibrosis.

The various symptoms of cystic fibrosis can be traced back to a single consequence of the faulty gene – the excessive production of abnormally thick and sticky mucus. In the lungs, this thick mucus clogs up the delicate alveoli (air sacs) and smaller bronchioles. As a result, breathing is difficult and the person is prone to lung infections. In the pancreas the large amounts of sticky mucus block the exit for the various pancreatic enzymes that normally go to our small intestines and help digest our food. As a result, a person with cystic fibrosis suffers from poor digestion as the enzymes fail to reach the small intestine. In addition, these pancreatic digestive enzymes – unable to reach the gut and accumulating in the pancreas – may start to attack the pancreas itself. As a conse-quence, the pancreas may literally be eaten away. As the pancreas produces the hormone insulin, which regulates blood sugar levels, people with cystic fibrosis may develop diabetes, which results from insufficient production of insulin.

In 1989, the defective gene responsible for cystic fibrosis was finally isolated.[15] The gene codes for a protein that controls the movement of chloride ions across cell membranes. The defective gene prevents the movement of chloride ions out of cells. Normally water follows these chloride ions and makes the mucus secreted by cells quite runny. In the absence of these chloride ions, less water accompanies the mucus, which is consequently more sticky. This simple fact is sufficient to cause all the problems associated with cystic fibrosis.

The isolation of the gene responsible for cystic fibrosis has opened up the reality of genetic screening. A simple test of a person's saliva, costing about the price of a newspaper, allows identification of carriers (people who have one healthy and one faulty copy of the gene), though only with about 90% accuracy. Suppose you have a close relative who has cystic fibrosis, but you don't. There is a high chance that you are a carrier. You might choose to undergo screening, for example if you are about to start a family.

Suppose you and your partner both discover that you are carriers, and the two of you are expecting a baby. As we have already said, there is a one in four chance that the baby will have cystic fibrosis. Genetic screening gives you the opportunity to use antenatal diagnosis to see if it will have cystic fibrosis or not. Using amniocentesis or chorionic villus sampling, a tiny sample of cells from the developing fetus is taken. (Both amniocentesis and chorionic villus sampling carry a small risk, of the order of 0.5 to 2%, of causing a miscarriage.) The relevant gene is examined in the laboratory and a diagnosis can be made. Three-quarters of the time the news will be good: the baby will not have cystic fibrosis. However, on a quarter of the occasions, the news will be bad: the baby will have cystic fibrosis.

At present the only option a couple in this position has is to decide whether to continue with the pregnancy or opt for a termination. Most opt for a termination, a decision of great personal and moral significance, particularly as most people with cystic fibrosis already live to be over 30. Gene therapy and conventional treatment are beginning to open up new avenues of treatment.

Somatic gene therapy for cystic fibrosis

As we have seen, cystic fibrosis causes problems for a person's lungs and for their digestive system. The problems with the digestive system can be controlled fairly well with drugs. However, the damage to the lungs eventually proves fatal and few people with cystic fibrosis live to be 40. Gene therapy offers the hope of halting, or at least slowing, the lung damage. The main steps are as follows:

1. Obtain a healthy copy of the CF (cystic fibrosis) gene.
2. Insert it into the genetic material of a convenient bacterium.
3. Allow the bacteria to reproduce many times.
4. Remove the healthy copies of the CF gene from the bacteria.

5. Put these healthy CF genes into a vector (such as a harmless virus).
6. Use this vector to carry the healthy CF genes to the cells that line the lungs.
7. Here the CF genes insert themselves into the DNA in these cells.
8. The CF genes then make the missing protein.
9. The missing protein then moves to the membrane that surrounds the cell.
10. Here it regulates the passage of chloride ions, allowing the mucus produced by the cell to be its normal runny consistency.

Three main vectors have been tried to allow step 6 to take place.[15,16] One approach is to use hollow membranous spheres called liposomes. The advantage of liposomes is that they easily fuse with the membranes that surround cells, thus releasing their contents into the cytoplasm of the cells. The disadvantage is that most of the genes carried in this way don't end up inserting themselves into the cell's DNA. A second approach is to use a retrovirus. Retroviruses are viruses that are very efficient at inserting genes into the cell's DNA. There is, though, a worry that there may be an increased risk of this damaging the normal controls on cell growth. In the worst case, this could conceivably trigger a cancer. The third approach is to use a cold virus. Cold viruses, of course, are rather good at carrying genes into lung cells. Unfortunately, they can also cause inflammation of the lungs – the last thing someone with damaged lungs wants.

The liposome approach was pioneered by Duncan Geddes at the Royal Brompton Hospital in London, Bob Williamson at St Mary's Hospital Medical School in London and a large research group at the University of Cambridge. The first trials on humans involved working with volunteers who had cystic fibrosis. Liposomes containing copies of the healthy CF gene were sprayed into their noses. Early indications have been encouraging. In particular, it has now been found that only 10% of the cells lining the lungs need to have the missing proteins in their membranes replaced for the lung to function normally.[17]

Alternatives to gene therapy for treating cystic fibrosis

Before we leave cystic fibrosis, we should note that even without gene therapy tremendous advances have been made in the treatment of cystic fibrosis. Average life expectancy for people with cystic fibrosis has increased hugely in recent years – in the UK from just one year in 1960 to five years in 1970 to ten years in 1980 to almost 20 years in 1990 and to around 25 years in 1995. These improvements have come about partly through aggressive antibiotic regimens, used to fight off lung infections, and partly through intensive, daily physiotherapy, which is time consuming and can be painful.

New advances continue to be made. In 1993 Genentech published the results of clinical trials on a new drug called pulmozyme. This reduces lung infection and shortness of breath.[18] In 1995, a Cleveland team published the results of a four-year clinical trial of a new drug called ibuprofen. They found that it slows lung deterioration by almost 90% in children who start taking it before their teens.[19]

Advances, such as these, in conventional medicine do not of course eliminate the potential of gene therapy. However, they remind us that gene therapy may not be the only way of treating even genetic diseases such as cystic fibrosis. In addition, the present work with liposomes and viruses as vectors should not be seen as a permanent solution to the problems of cystic fibrosis. Because the cells lining the lungs are shed quite rapidly, repeated gene therapy will probably be needed. Nevertheless, there is a real chance that gene therapy may come to play a significant part in the treatment of cystic fibrosis, allowing tens of thousands of people to enjoy a far better quality of life.

Somatic gene therapy for other diseases

We have concentrated so far on cystic fibrosis because of its widespread occurrence and because of the potential genetic engineering has to help people with this condition. However, the first successful attempts to genetically engineer humans were carried out in 1990, three years before trials began on cystic fibrosis.

These attempts involved patients with a very rare disorder known as severe combined immune deficiency (SCID). In someone with SCID, the immune system doesn't work. As a result the person is highly susceptible to infections. Children with SCID are sometimes known as bubble babies because, until recently, almost the only way to allow them to live for more than a few years was to isolate them in plastic bubbles. These bubbles protect the children from harmful germs but also, poignantly, cut them off from all social contact. In any event, at best the bubbles prolong life by only a few years.

SCID can have a number of causes. Probably the single most common one is an inherited deficiency in a single enzyme, adenosine deaminase (ADA). The first person with SCID to be treated with gene therapy was a four-year-old girl called Ashanthi De Silva. She was unable to produce ADA and in 1990 some of her white blood cells were removed, and functioning versions of the ADA gene introduced into them using a virus as a vector.[20] The improvement in her condition was remarkable. Five years on she was living a comparatively normal life, attending a typical school and so on. She was no more likely to catch infections than her classmates and, on one notable occasion, when she and all her family caught flu, she was the first to recover. At present she needs regular transfusions (every few months) of genetically engineered white blood cells as the white blood cells live less than a year. It may eventually be possible to modify so-called stem cells. Stem cells are immortal; they reside in the bone marrow and give rise to the different types of blood cell. At present, though, identifying and purifying stem cells is proving difficult. If they can be genetically engineered they hold out the hope of a complete cure for SCID (Figure 8.1).

By 1995, over 200 trials for somatic gene therapy had been approved. In addition to trials on people with cystic fibrosis and SCID, somatic gene therapy is being tried for a number of other conditions including familial hypercholesterolaemia, Duchenne muscular dystrophy, haemophilia, β-thalassaemia and cancers. In many cases, the results have been very encouraging.[17,21]

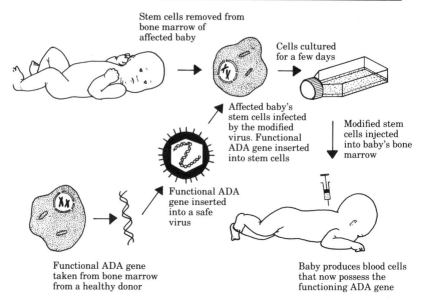

Stem cells removed from
bone marrow of
affected baby

Cells cultured
for a few days

Affected baby's
stem cells infected
by the modified
virus. Functional
ADA gene inserted
into stem cells

Modified stem
cells injected
into baby's bone
marrow

Functional ADA
gene inserted
into a safe
virus

Functional ADA gene
taken from bone marrow
from a healthy donor

Baby produces blood cells
that now possess the
functioning ADA gene

Figure 8.1. An attempt at a permanent cure for severe combined immunodeficiency (SCID) by genetically engineering bone marrow stem cells. This procedure was first carried out at Great Ormond Street Hospital, London in 1993.

The scope of somatic gene therapy

It is easy, given the initial successes of somatic gene therapy, to get carried away, hoping that it will soon be a cure for all our problems (Figure 8.2). It is important, however, to be clear about the potential scope of gene therapy. For a start, it should be realised that some human diseases caused by faulty genes can already be treated quite effectively by conventional means. For example, every baby born in the UK, and in a number of other countries, is tested for the genetic disease phenylketonuria, though the baby's parents probably aren't aware of this. The reason is that, provided action is taken soon after birth, the harmful consequences of the condition can be prevented.[22] Phenylketonuria is a condition which, if untreated, leads to the person being severely mentally retarded. Affected individuals often have convulsions and, in the past, were frequently institutionalised. Since 1954, though, it has

Figure 8.2. Some people have unrealistic expectations of genetic engineering.

been realised that the condition can be entirely prevented. If children with the faulty gene that causes phenylketonuria are given a diet that has only small amounts of the amino acid phenylalanine, they grow up healthy and normal. This is because they lack the particular enzyme that converts phenylalanine into another amino acid, tyrosine. It is the build-up in the levels of phenylalanine that causes the problems. Keeping the levels of phenylalanine in the diet low thus prevents the harmful consequences of the condition.

Phenylketonuria illustrates a most important truth about human development: both genes and the environment play essential parts. Yes, phenylketonuria is a genetic disease in the sense that it is the result of a faulty gene. But the extent to which the disease manifests itself depends on the environment. A normal diet (i.e. normal environment) and the person is severely affected; a special diet (i.e. a different environment) and the person is unaffected. To describe phenylketonuria, or any other condition, as a 'genetic disease' is, at best, a convenient shorthand. Both genes and the environment are involved in the manifestation of any trait. Shorthand is useful so long as one doesn't forget what it stands for.

So changing the environment can prevent some genetic diseases. In essence, changing the environment is how all conventional medicine works. Take juvenile-onset diabetes, which we looked at earlier (p. 97). Again, the disease is 'genetic'. In this case, though, its symptoms can largely – though not entirely – be prevented by regular injections with insulin. As we saw, this insulin can be obtained from pigs or cows, or it can be produced by bacteria that have been genetically engineered. Of course, it may turn out to be the case that the best results will come from genetically engineering humans so that their pancreases produce their own insulin 24 hours a day. However, the fact that millions of diabetics have been able to lead relatively normal lives thanks to insulin injections, instead of dying painfully in childhood or adolescence, shows how both genes and the environment play a vital role in development.

A second reason why we should not see gene therapy as the likely solution to all medical problems also relates to the roles played by genes and the environment. Frequent announcements in the press that the gene for breast cancer, cancer of the colon, Alzheimer's disease, schizophrenia or whatever has been identified may appear to offer the hope for a cure.[23] The reality, though, is often far from this.

First of all, knowing what causes a condition may be a valuable step in preventing it, but it is most definitely not the same thing. You may be unable to speak Russian or play the trumpet, and, if this is the case, you probably know why you can't. But knowing this is no more than the first step in helping you achieve these ends. In the same way, knowing which of the 80000 or so human genes causes a disease is a very long way from treating, curing or preventing it. Initially it simply offers the possibility of genetic screening.

Secondly, diseases such as cystic fibrosis, phenylketonuria and sickle-cell disease are the exception, not the rule. These conditions are caused by inborn errors in single genes. However, only 2% of our total disease load results from either errors in single genes, such as these, or chromosome mutations such as those leading to Down's syndrome and Turner's syndrome, where the person is born with the wrong number of chromosomes. Around 98% of human disease is not like this. For a start, most human diseases have a strong environmental component, so that genetic defects merely predispose the person to develop the condition. Then, the genetic component is the result of many genes.[24]

A condition affected by many genes is called a 'polygenic' condition or trait. Human skin colour, for instance, is a polygenic trait. At school, pupils are taught a very simplified, and somewhat misleading, version of human genetics. For example, human eye colour is generally taught as being determined by just a single gene. In reality, though, eye colour is determined by several genes. Indeed, the vast majority of human traits are determined polygenically.

A final point is that we are only just beginning to appreciate the extent to which the human body, and mind, can overcome genetic handicaps. Particular mutations certainly increase the risk of an individual developing certain cancers or heart disease. Similarly, there may well be mutations that increase one's risk of developing schizophrenia or depression. However, this is a million miles away from saying that there is 'a gene for breast cancer' or 'a gene for schizophrenia'. In the case of breast cancer, early indications are that while a mutation in the BRCA1 gene greatly increases a woman's chance of developing breast cancer, it is still the case that at least 15% of women with this mutation do not go on to develop breast cancer. Further, it is probable that many women with a history of breast cancer in their family do not have a mutation in this gene.[25] In the case of the 'colon gene', a mutation in the gene accounts for only 10% of all colon cancers, while someone with an altered form of the gene has a 30% chance of never developing colon cancer.[26]

The fact that most important human diseases have a heavy environmental component, are almost always affected by several genes, rather than just one, and are frequently modulated in their severity by the actions of the human body means that the scope for gene therapy is likely to be more limited than many suspect.

Somatic gene therapy for other human traits

So far we have confined ourselves to what we may term 'real human diseases'. But what of gene therapy to affect traits such as intelligence, beauty, criminality and sexual preference. Will this ever be practicable?

Although there are frequently reports in the popular press of 'a gene for homosexuality' or 'a gene for criminality', our discussion of the complexity of human disease should caution us against such simplistic notions. Yes, it may be that as we learn more about our

DNA we will find out that there is a genetic component to many human traits. Certainly much dog behaviour has a significant genetic component.[27] If you present a Border collie puppy (bred to round-up sheep) and a Newfoundland puppy (bred to rescue drowning sailors) with, for the first times in their lives, a collection of scattered tennis balls and a large bowl of water, the Border collie will often attempt to round up the tennis balls and the Newfoundland will jump into the water.

In the same way, much human behaviour almost certainly has a genetic component to it. However, attempts to find genes for homosexuality, intelligence, beauty or criminality are, at best, the first steps to understanding the rich and complex ways in which we behave.[28] At worst, they are misguided attempts to stigmatise certain members of society (Figure 8.3). Part of the very essence of our being human is that we, above all the other organisms with which we share this planet, have the potential to transcend much of our biological heritage. We are more, far, far more than our genes.[30]

The ethical significance of somatic gene therapy

What new ethical issues are raised by somatic gene therapy? The short answer, when we are talking about real human diseases, is probably none. Of course, somatic gene therapy is still a very new technique, and mostly it is at the research rather than the clinical stage. However, there is considerable agreement about how medical research and innovative practice should be regulated in the light of ethical considerations.[31,32] There is little doubt that ethical considerations have so far been applied to somatic gene therapy even more stringently than to conventional medicine.

In the UK, the Clothier Committee produced its report on the ethics of somatic gene therapy in January 1992.[32] After carefully reviewing such considerations as the likely success of the procedure, consent and confidentiality, the Clothier Committee's decision was that:

> We conclude that the development and introduction of safe
> and effective means of somatic cell gene modification,

Figure 8.3. In 1992, a conference on genetics and crime was cancelled amid angry accusations and counter-accusations.[29] The conference called 'Genetic Factors in Crime' was planned for October 1992. A brochure introducing the conference stated that 'genetic research holds out the prospect of identifying individuals who may be predisposed to certain kinds of criminal conduct'. A key figure opposing the conference was psychologist Peter Breggin of Bethesda, Maryland. Breggin maintained that the conference was part of an initiative planned by the National Institute of Mental Health in which children in inner-city schools would be tested for biological markers, such as low levels of 5-hydroxytryptamine, that allegedly make people more likely to be violent. Most American inner-city schoolchildren are black.

directed to alleviating disease in individual patients, is a proper goal for medical science. Somatic cell gene therapy should be regarded, at first, as research involving humans subjects and we **recommend** that its use be conditional upon scientific, medical and ethical review. Although the prospect of this new therapy heightens the familiar ethical concerns which attend the introduction of any new treatment, we conclude that it poses no new ethical problems.[33]

Although the UK government waited over a year to respond, it eventually accepted this recommendation.[34] Other countries too have permitted somatic gene therapy. Because somatic gene therapy typically involves giving a person healthy DNA to override the effects of their own malfunctioning DNA, it has been pointed out that this is not very different from giving a person a blood transfusion or organ transplant. Of course, some individuals may choose not to have a transfusion or transplant, but very few people suggest forbidding them entirely.

It is also the case that somatic gene therapy has the potential to reduce the number of ethically problematic decisions. We have already mentioned (p. 204) how at present the only 'solution' offered to a woman who is carrying a fetus identified as having a serious genetic disorder such as cystic fibrosis or muscular dystrophy is the possibility of an abortion. Somatic gene therapy may be able to offer a more positive way forward. Additionally, in some countries fetal tissue obtained from abortions is being used in research connected with a number of diseases, particularly Parkinson's disease.[35] Work by Olle Lidvall in Sweden has shown what have been described as 'Astonishing improvements in patients given implants of fetal brain tissue to relieve symptoms of Parkinson's disease'.[36] There have even been calls from a professor of medicine for abortions to be carried out more carefully to ensure that the fetus emerges alive so that its organs can be used for medical purposes.[37]

Somatic gene therapy offers the hope of a different approach. Ning-Sun Yang of Agracetus, a biotechnology company in Middleton, Wisconsin, is trying to genetically engineer rats (work that in its use of sentient animals is itself ethically problematic, as we discussed in Chapter 7) to produce dopamine. Dopamine is a chemical produced by the brain but which people with Parkinson's disease lose the ability to make.[38] The research is still at a very early stage but, if successful, might eliminate the use of aborted fetuses for the treatment of Parkinson's disease. Aside from the ethical considerations, the use of aborted fetuses for subsequent surgery is expensive, around $40000 in the USA, and not very efficient – only 5 to 10% of the fetal cells survive the graft.[39]

However, somatic gene therapy may, in time, raise new ethical issues. Suppose, despite what we have said about the complexities of human behaviour, it does eventually transpire that somatic gene therapy could reduce the likelihood of someone being violently aggressive or of being sexually attracted to others of the same sex. What then? The simple answer is to throw one's hands up in horror and agree that such 'treatments' should be outlawed. However, one problem with this response is that most countries already spend a lot of time and effort trying to get people who have been convicted of violent crimes to be less likely to commit these again. They may attend education programmes or receive state-funded psychotherapy, for instance, in attempts to achieve these aims. Similarly, some psychiatrists and counsellors are still prepared to work with homosexuals to help them change their sexual orientation.

These two examples (violent behaviour and homosexuality) highlight two related issues. The first is to do with the social construction of disease. It is tempting to think that a disease is a disease, full stop. The reality is that a disease is, in a sense, a relationship a person has with society. As we asked in Chapter 5, is being 4 foot tall a disease? The answer tells us more about a society than it does about an individual of this height. Some conditions are relatively unproblematic in their definition as a disease. For instance, Lesch–Nyhan disease is characterised by severe mental retardation, uncontrolled movements (spasticity) and self-mutilation. No cure is at present available and the person dies, early in life, after what most people would consider an unpleasant existence. It is the existence of conditions such as this that have even led to claims in the courts of wrongful life or wrongful birth where a sufferer, or someone acting on their behalf, sues either their parent(s) or doctor(s) on the grounds that it would have been better for them never to have been born. However, years of campaigning by activists for people with disabilities have shown us the extent to which many diseases or disabilities are as much a reflection of the society in which the person lives as they are the product of the genes and internal environment of that person.

The second issue is to do with consent. It is one thing for a person convicted of a violent crime to give their informed consent to receive psychotherapy or some other treatment aimed at changing their behaviour – though even these treatments are, of course, open to abuse.[40] It would be quite another for a parent to decide on a fetus' or baby's behalf to let it receive somatic gene therapy to make it less aggressive.

In an attempt to set limits on the operation of somatic gene therapy, the Clothier Committee concluded that: 'In the present state of knowledge any attempt by gene modification to change human traits not associated with disease would not be acceptable'.[41] Similarly, the arguments of members of a number of the world's major religions at the 1992 conference on 'Genetics, Religion and Ethics' held in Houston, Texas, are fairly represented by the view of George Pazin, a member of the Orthodox Church and professor at the University of Pittsburgh School of Medicine: 'I am all in favour of repairing God's creation with the genetic tools that we have discovered, but I shudder to think of our trying to improve upon the creation'.[42]

At the present time, it may be difficult to be more precise than these last two quotes. However, it is worth noting that several countries officially permit abortions only on health grounds yet, in practice, offer abortion on demand. By analogy, should the use of somatic gene therapy ever become widespread it may be difficult to prevent it being used for cosmetic purposes, in much the same way that plastic surgery can be used both for life-saving and for trivial purposes.

Germ-line therapy

The idea of genetic alterations to the human germ-line (so that succeeding generations are affected, rather than just the individual concerned) has been rejected by a number of religious writers and organisations,[43] as it has by many secular writers and organisations.[44] The main arguments against human germ-line therapy are as follows:

- It is too risky
- It is unnecessary
- It is wrong.

However, others, including a number of distinguished moral philosophers and some theologians, have argued that the time may come when germ-line therapy is permissible, even highly desirable.[45] Some people believe that time is fast approaching. In 1994, the University of Pennsylvania in Philadelphia filed a patent application for a technique devised by Ralph Brinster and Jim Zimmermann at its veterinary school.[46] Brinster and Zimmermann developed a technique for removing the immature reproductive cells found in a mammal's testes, genetically modifying them and then replacing them in the testes. The patent application specifically mentions humans as well as other mammals. Brinster has been reported (p. 5) as saying that he would be surprised if the patent caused a moral backlash: 'I had not thought about it, because I think about it as potentially benefiting man'.[46]

We shall look in turn at the three main objections raised against germ-line therapy in humans, keeping in mind that, as we have seen, germ-line alterations are already well established for microorganisms (Chapter 5), plants (Chapter 6) and animals (Chapter 7).

Is germ-line therapy too risky?

At the moment it is generally acknowledged that human germ-line therapy is too risky. Researchers cannot, at present, control precisely where new genes are inserted. This raises the not insignificant danger that the inserted gene might damage an existing gene, which could lead to diseases, including cancers. We can note, in passing, that the existence, despite these problems, of germ-line therapy in animals (i.e. non-human animals) illustrates the distinction between what is generally deemed acceptable for animals and for humans.

However, although human germ-line therapy may currently be too risky, it is difficult to imagine that this will continue to be the case indefinitely. It seems extremely likely that scientists will develop methods of targeting the insertion of new genes with sufficient precision to avoid the problems that presently attend such procedures.[47] Nor need these new methods require much, possibly any, experimentation on human embryos. A great deal, perhaps all, of the information could be obtained through the genetic engineering of farm animals.

Further, we should realise that although germ-line therapy is sometimes referred to as 'irrevocable', it is more likely, if we ever get to the point where its use is routine, that it will normally be reversible. There is no reason to suppose that if something went wrong with the results of germ-line therapy, this wrong would necessarily be visited on a person's descendants in perpetuity. The same techniques that will permit targeted germ-line therapy should permit its reversal.

Is germ-line therapy unnecessary?

As we have already discussed (p. 125), it is no easy matter to demonstrate that something is 'necessary'. Value judgements are involved, so that there may be genuine controversy about whether something is needed. Is nuclear power necessary? Or the motor car? Or tigers? Or confidentiality between doctors and their patients? It is likely that most improvements that might result from germ-line therapy could also be effected by somatic gene therapy or conventional medicine. Cystic fibrosis and diabetes are cases in point. However, it may prove to be the case that germ-line therapy allows such conditions to be treated better. It is possible that germ-line therapy might be able to produce certain benefits that could not be realised by any other technique. No doubt the human race would be able to get on without germ-line therapy, and one may question whether it would do much to increase the sum of human happiness. Nevertheless, at some point it may convincingly be argued that germ-line therapy is necessary.

Is germ-line therapy wrong?

Making the assumption, then, that one day germ-line therapy will be both relatively safe and deemed necessary, in the sense that it can bring benefits that other approaches can't, is it right or is it wrong?

It is sometimes argued that germ-line therapy will decrease the amount of genetic variation among people and that this is not a good thing, since evolution needs genetic variation. There are several things that are dubious about this objection. First, empirically, it is difficult to imagine in the foreseeable future that germ-line therapy is going significantly to decrease the amount of useful human variation among people. Secondly, it is possible that germ-line therapy may one day lead to even more genetic variation – as some parents opt for certain genes in their children and other parents for other genes. Thirdly, the argument that evolution needs genetic variation is difficult to sustain faced with someone suffering as a result of a disease that is largely the result of a genetic mutation. The argument relies on possible, very distant advantages for groups of people being sufficient to over-ride the more immediate, clear disadvantages for individuals.

Then some people have expressed the fear that germ-line therapy might be used by dictators to produce only certain types of person. The emotive term 'eugenics' is often used in this context. Perhaps the major problem with this objection is that it assumes too much of genetic engineering. As we discussed in the context of somatic gene therapy, it is easy to overstate the extent to which humans are controlled by their genes. Dictators have had, have now and will have far more effective ways of controlling people.

A more likely problem is that germ-line therapy will be permitted before people have grown sufficiently accustomed to the idea. The pace of technological change is so fast nowadays that some people end up feeling bewildered by new possibilities. The theologian Ian Barbour has argued that it is important that sufficient time is allowed before germ-line therapy on humans is permitted, both to ascertain, so far as is possible, that the procedure is safe and so that people may feel comfortable with the idea:

> . . . I would approve germ-line therapy only under three conditions. First, extensive studies of human *somatic-cell* therapies similar to the proposed germ-line therapy must have been conducted over a period of many years to acquire data on the indirect effects of the genetic changes. Secondly, the effects of similar germ-line therapy in *animals* must have been followed over a period of several generations to ensure the reliability and long-term safety of the techniques used. Third, widespread *public approval* must have been secured, since the therapy will affect unborn generations who cannot themselves give informed consent to treatment.[48]

A frequently expressed worry about germ-line therapy is the extent to which future generations will be affected. Again, it is possible that this fear may be an exaggerated one. As we have said, we can overestimate the importance of our genetic make-up. Then there is the point that people already have and will continue to have a tremendous influence over future generations through everything from childrearing patterns and family planning to books and pollution. The philosophers Robert Nozick,[48] Jonathan Glover[45] and John Harris[50,51] have been quite bullish about germ-line therapy and Nozick, back in 1974, introduced the notion of a 'genetic supermarket' at which parents, rather than the state, could choose the genetic make-up of their children. Harris has even argued that we may have a duty to carry out germ-line therapy:

> We must not act positively so as to cause harm to those who come after us, but we must also not fail to remove dangers which, if left in place, will cause harm to future people. Thought of in this light, there is a clear dilemma about genetic engineering. On the one hand we must not make changes to the genetic structure of persons which will adversely affect their descendants. On the other hand we must not fail to remove genetic damage which we could remove and which, if left in place, will cause harm to future people.[52]

There remains the worry, though, born of long experience of slippery slopes, that the road to hell is paved with good intentions. Despite the difficulties, which we reviewed earlier, of distinguishing in all cases genetic engineering to correct faults (such as cystic fibrosis, haemophilia or cancers) from genetic engineering to enhance traits (such as intelligence, creativity, athletic prowess or musical ability), the best way forward may be to ban germ-line therapy intended only to enhance traits, at least until many years of informed debate have taken place.

The idea that it is useful to distinguish between genetic engineering to treat disease and genetic engineering to enhance traits was met in Chapter 4 when we saw that some theologians have argued that genetic engineering to treat disease may be seen as part of the redemptive activity of humans. At the same time, it is worth noting the caution of J. Robert Nelson, Senior Research Fellow of

The Institute of Religion, Texas Medical Centre and Adjunct Professor of Medical Ethics at Baylor College of Medicine: 'The prospect of overcoming and even eliminating from the germ-line certain types of human suffering is, like all other eschatologies, both appealing and frightening'.[53]

John Habgood, the former Archbishop of York, has cautioned against using genetic engineering to improve people. In a talk given in 1995 he concluded:

> So, six rules in a sentence. First, human beings are more than
> their genes. Genes are only a set of instructions. We are more
> than a set of instructions. Second rule: remember the valuable
> diversity of human nature. Third rule: look for justice in the
> dealings of human beings with one another and for fairness
> in the use of resources. Fourth rule: respect privacy and
> autonomy. Fifth rule: accept the presumption that diseases
> should be cured when it is possible to do so, And sixth rule:
> be very suspicious about improving human nature; and be
> even more suspicious of those who think they know what
> improvements ought to be made.[54]

Nor is it only religious leaders who have warned of the dangers of presuming to improve human nature or enhance human capabilities. Jonathan Glover quotes the philosopher John Mackie who once argued, against Glover's optimism about germ-line therapy, that 'if the Victorians had been able to use genetic engineering, they would have aimed to make us more pious and patriotic'.[55]

A related difficulty is that if we ever did succeed, through either somatic cell or germ-line therapy, in enhancing such traits as intelligence, memory or altruism, there might be unforeseen consequences. There exists a genetically engineered strain of the fruit fly *Drosophila* that learns ten times faster than the normal strain.[56] At first sight, the application of this technology to humans sounds marvellous. Imagine learning ten times faster; think of all the benefits it could bring. However, there may be costs. Improved learning implies improved memory and if you have a far superior memory you will forget far less. Most of us have experienced unpleasant happenings that we are only too grateful to forget. Further, there is some anecdotal evidence to suggest that the handful of people who have total recall or perfect photographic memories find life difficult. For a start, they don't always find it easy to know what day, month or even year it is. If you have a perfect memory, events a year ago may be almost as fresh in your mind as events five minutes ago. This can lead to confusion and can make social relationships difficult.

Finally, it might be the case that genetic engineering would require parents to choose which traits they would like enhanced in their children. It may, for example, prove impossible simultaneously to enhance a child's ability to learn mathematics, paint, show great empathy and play a musical instrument. It can be argued that genetic engineering to enhance human traits diminishes the autonomy of the person concerned, i.e. the genetically engineered child that results. On a more practical level, one can imagine arguments between genetically engineered children and their parents, with great unhappiness sometimes resulting. 'We didn't pay for you to be musically gifted just to have you spend all your time playing baseball' or 'We didn't pay for you to be an outstanding baseball player just to have you spend all your time in a rock band', etc.

Conclusions

Somatic gene therapy for a number of medical conditions, such as cystic fibrosis, is beginning to offer the realistic possibility of improving physical health, relieving stress, worry and other psychological problems, and reducing the number of abortions. Further it is not significantly different, in principle, from other medical treatments. Those who accept conventional medicine – well over 99.9% of all people – should have no qualms about accepting the benefits of somatic gene therapy provided it is shown to be safe and efficacious.

However, the likely benefits of somatic gene therapy should not be overstated. Most human disease will not be eliminated through its adoption. In time somatic gene therapy is likely to prove to be just one of a number of important therapeutic approaches used in the prevention and treatment of human disease.

At present, germ-line therapy is unsafe. In time, this may change. In our opinion, the ethical arguments in favour of a total, unconditional and permanent ban on germ-line therapy are not overwhelming. There may prove to be strong arguments in favour of the careful use of germ-line therapy to alleviate suffering in certain circumstances.

For the foreseeable future, it is probably best to outlaw the use of either somatic or germ-line therapy to enhance human traits. In time, such applications of genetic engineering may be practicable and safe. However, not everyone will agree on what changes, if any, should be made. In such circumstances it would be better to err on the side of caution.

PART 3

9

Public understanding of genetic engineering: what can education do?

Introduction

We have seen that genetic engineering raises many issues that are complex and multidimensional. These dimensions include the scientific and technological, the ethical and theological, and the economic and political. Given this range of perspectives, it is not surprising that controversy is more common than consensus in this area. Yet on one point widespread agreement does appear to exist. Virtually any conference, workshop or debate about genetic engineering can be safely predicted to end with a concerted call for 'more education' and for 'greater public understanding'. Scientists, politicians, industrialists, academics and consumer representatives speak largely with one voice in claiming that everyone needs to know *more* about genetic engineering, and this book is itself an attempt to contribute to this educational process.

The objective of increasing public awareness and understanding about genetic engineering, however, is by no means as straightforward and uncontroversial as it may at first appear – indeed, it raises a whole set of *further* questions, including some difficult ethical and educational ones. This chapter will, therefore, examine some of these crucial questions, which tend to be overlooked by those who agree that 'what we need is more education' but who may not have thought out the ethical and educational implications of this claim.

Why does the public need to know more?

Although an apparent consensus exists here, this disguises the fact that those who call for more education and greater public understanding do so for a wide variety of different and at times conflicting reasons, which reflect very diverse attitudes, interests and assumptions. To illustrate this, the following sample of viewpoints can be identified:

The 'conspiracy theory' view

According to this, genetic engineering and its applications are highly suspect on all kinds of grounds and likely to benefit only a narrow range of commercial and political interests. Extreme versions of this view can suggest that the public is being kept in the dark about these negative features and that those with a vested interest in the development of the technology want to maintain this veil of ignorance. Therefore, to counter this policy the public needs to be told much more about genetic engineering and so awakened to the dangers. The assumption behind this view, then, is that better understanding of genetic engineering will lead the public to be very wary of it and probably reject it.

The 'optimistic' view

This is diametrically opposed to the conspiracy theory view in that it believes that better understanding will lead to public *acceptance* rather than rejection. Public acceptance is deemed important in order that what are seen as the enormous potential benefits of genetic engineering can come to fruition. The great fear of the many industrial and political proponents of this view is that huge investments in this new technology may be wasted if the public finally rejects its products, and the main reason why this might happen, according to the optimists, is that the public does not know enough about the technology to appreciate its likely benefits.

The 'democratic' view

This is a more neutral position than either of the above, in that it does not have either acceptance or rejection as its direct objective. An increasing number of policy decisions, it is argued, will have to be made about applications of genetic engineering and as these are likely to affect all of us in various ways we all need to be able to participate somehow in this decision-making or at least to be aware of the implications for us. The assumption here is that these issues cannot just be 'left to the experts' (e.g. scientists, industrialists and politicians); public opinion must also influence the decision-making, but it must be *well-informed* public opinion, and that requires education.

The 'consumer choice' view

This is a more specific and individually focused version of the democratic view. Members of the public as individual consumers, according to this view, will be faced with a growing number of products and options as a result of genetic engineering. If they are to be able in this situation to exercise a free and informed choice, which has variously been described as 'the most precious of all consumers' rights'[1] and as 'the engine of consumer power',[2] they need to understand the distinctive features of those products that involve genetic engineering to enable them to assess their relative advantages and disadvantages.

The 'survival' view

This is the most general and perhaps least contentious approach, claiming simply that we all need to know more about genetic engineering because it will inevitably play an increasingly important part in all of our lives. We, therefore, require education in this area as in many others, to enable us to 'cope' with the various ways in which it may impinge upon our lives. Similar arguments can be forwarded in the case of other modern technologies, for example information technology: we need to know about such developments in order to survive in a society that is likely to be significantly influenced and shaped by them.

No doubt other views or variants of the above five could also be identified, but this list is sufficient to show that those who agree about the need for greater public understanding do not necessarily agree about much else. Apart from the obvious diversity of 'political' viewpoints represented above, there are also some interesting factual disagreements implicit in some of them. If people learn more about genetic engineering and its applications, for example, will that lead them to reject it as the advocates of the conspiracy theory assume, or accept it as the optimists believe? What evidence we have at present does not indicate a clear-cut answer,[3] but the question itself raises important educational issues to which we will return later in this chapter.

Education or regulation?

Despite the large measure of agreement that appears to exist in favour of increasing the public's awareness and understanding of genetic engineering, an alternative line of argument is sometimes voiced that needs to be considered at this point. According to this viewpoint, the public does *not* need to know more about such highly technical subjects, and indeed it is unrealistic to expect that most members of the public, lacking a formal scientific background, would want or be able to gain understanding of matters that are located on the frontiers of modern scientific research. What they *do* want and need, the argument goes, is confidence and trust in the regulatory authorities to ensure that genetic engineering and its varied applications are safe and strictly monitored. If, for example, a new genetically engineered food product becomes available, the public does not need to understand the technology that has produced it, but it does need to know that independent experts are sure that it is safe to eat that product; market forces can then be relied on to determine whether or not the product is a commercial success.

Although this argument has a certain commonsense plausibility about it, on closer inspection it reveals a number of flaws that relate to points which have already been discussed in Chapter 3. The notion of 'safety' was shown in that chapter to be less clearcut than is generally supposed. No process or product can ever be guaranteed 100% 'safe', for no amount of current data available today can prove that certain events will or will not happen in the future; further, new knowledge and evidence can quickly reverse scientific judgments of what is safe and what is dangerous. So, strictly speaking, no 'expert' can or should ever be expected to guarantee the safety of anything. This means that practical decisions about safety are not purely factual or empirical, as they must also involve the weighing of possible risks and benefits, which in turn necessarily involves the making of *value judgments*.

Now while it might be thought reasonable for the public to 'trust the experts' on purely scientific matters, it is far less obvious that that trust should extend into the area of values. Should we hand over to 'experts' the right to make value judgments for us, and if so who would these experts be? Scientific experts have no special right or qualification to pronounce upon moral values, and indeed it is debatable whether *anyone* could or should ever claim 'moral expertise' to be exercised on behalf of the general public. Moral judgments cannot be proved or demonstrated to be unequivocally right or wrong, and many would argue that the only person who can make a genuine moral decision for me is myself – though that is not to say that all moral decisions are necessarily equally justifiable or well-reasoned.

Relying wholly upon experts to pronounce upon the safety of genetic engineering cannot, therefore, remove the need for education and better public understanding, because safety is not an entirely technical issue. Also, we have seen in the previous sections that safety is by no means the only aspect of genetic engineering capable of generating public concern. The fact that a particular product or process of genetic engineering is declared 'safe' by a panel of experts (despite the above objections to this notion) does not remove other possible legitimate reasons why people might feel concern about that product or process and believe that it ought not to be offered or developed.

This illustrates how the 'public understanding' of genetic engineering (or of any other controversial scientific development) is in fact an ambiguous phrase, as it can refer to knowledge of the purely scientific aspects of the subject or to awareness of its wide-ranging moral, religious, economic, social and political implications. While the former is clearly possible without the latter, the non-scientific implications of genetic engineering cannot really be appreciated without some minimal understanding of its scientific basis. If members of the public are to be able to make informed decisions about genetic engineering, they will need to have some 'understanding' of it in both the above senses.

What does the public already know?

Another way of countering the claim that the public needs to know more about genetic engineering might be to suggest that they know enough about it already. So what evidence do we have about the extent of that existing knowledge? Studies have been carried out in a number of countries to try to establish the current level of public understanding, though rather more emphasis has been placed upon discovering what people's *attitudes* are towards genetic engineering rather than what they actually *know* about it.

Despite these studies, it is extremely difficult, even if not impossible, to draw any general conclusions about the current level of public understanding for the following reasons:

1. 'Public understanding' is a highly amorphous term. Apart from the ambiguity just noted between the scientific and non-scientific aspects of this understanding, it begs the question of *which* public is being referred to. Obviously the situation will vary considerably from country to country and from continent to continent, but even within an individual country no single undifferentiated 'public' can be said to exist. People differ in terms of their age, sex, education, occupation and countless other respects, and these differences will significantly influence their fund of knowledge, their attitudes and their beliefs. The 'publics' in the UK, for example, who constitute the

readership of the Sun, the Guardian and the Financial
Times are unlikely to share much common
'understanding' of genetic engineering.

2. Developments in this field are taking place rapidly.
    Hardly a week goes by without the report of some new
    discovery or application appearing in the press. Public
    understanding cannot then be a static entity, as it is
    being constantly modified by new information. Shoppers
    in supermarkets, for example, are likely to encounter an
    increasing number of products that involve genetic
    engineering in some form. So any attempt at a
    'snapshot' assessment of the current level of public
    understanding will probably be out of date by the time
    its findings are disseminated; while trying to say
    anything meaningful on this subject in a book, which
    inevitably involves a delay between writing and
    publication and which will continue (we hope) to be read
    for several years, seems particularly hazardous.

Given these limitations, we shall not attempt here to generalise
about the current level of public understanding. Nevertheless,
some of the recent research done on this subject is both interesting
and important, and it deserves a brief mention at this point as a
response to the view that the public knows enough about genetic
engineering already.

As an overall generalisation, most of the large surveys show that
a high proportion of people are ignorant of key features of genetic
engineering. This is true both of adults and of school children.[4]
For example, questionnaires and interviews generally show that,
although the majority of people have heard of genetic engineering
or biotechnology, fewer than a quarter of us can give any sort of
adequate account of what these terms mean. Similarly, most
people are unable to explain the meaning of such essential words as
'DNA' or 'gene', nor are they able to identify more than one or two,
if that, uses or potential uses of the new technologies.

At first sight, this conclusion may appear somewhat alarming.
However, we should note that while few of us may be able to give
any sort of account as to what is meant by nuclear fission, this may

not prevent us from having quite well-informed views about the potential benefits and risks of nuclear power. There is always a danger that surveys about the 'public understanding of science' may focus on what is easy for a researcher to determine, rather than what is important. We need to enquire in more depth what it would mean to be 'more educated' about genetic engineering.

What does 'more education' mean?

Even if there is fairly general agreement that the public needs to know more about genetic engineering, considerable problems remain about what exactly this entails. In particular, the call for 'more education' in this area begs some important questions and blurs a number of crucial distinctions, which we need to clarify at this point:

Who is to be 'educated'?

Education is often assumed to refer to what is offered to young people in schools, colleges and universities, and one obvious long-term way of ensuring that the public understands more about genetic engineering is to devote more attention to all aspects of the subject in such educational institutions. Education cannot, however, simply be equated with schooling, for many things go on in schools that cannot count as education (or successful education), and conversely education frequently goes on outside schools. Adult education, for example, refers to educational experiences and activities that do not occur in schools. So while some educational initiatives in this area will obviously be school-based, school children and students form only a relatively small proportion of the 'public' who, it is claimed, need more education. Indeed, while school-based education may offer the most promising long-term solution, it cannot do much to address the more immediate problem of raising the understanding of the current adult population, who are already encountering various applications of the new technology.

Educating or informing?

It is clearly unrealistic to propose full-scale educational courses on genetic engineering for all members of the public; yet many would maintain that everyone needs to know *something* about the subject for one or more of the reasons elaborated earlier.

A distinction, then, needs to be drawn between educating and informing. Educating is a long-term enterprise, aiming at developing a broad-based knowledge and understanding in learners and giving them some grasp of the general principles underlying whatever is being learned. Informing is a much more limited, direct and specific imparting of factual material and is consequently a much less ambitious activity than educating. Consumers, for example, could be informed about the features of a particular genetically engineered product, but this would hardly constitute education – despite the grandiose references that are sometimes made to 'consumer education' when all that is usually meant is the giving of specific information.

Education, however, involves much more than the giving of information:

> Successful education must, by definition, affect how people
> think and judge and assess and deliberate, how they draw
> conclusions and make decisions, and how they act. This is
> achieved largely through the medium of reasons. The learner
> is subjected to various experiences which are designed to
> increase his or her knowledge, understanding, appreciation
> and sensitivity.[5]

So educating people about genetic engineering will not mean just telling them about particular processes and products; it will help them to understand that different sorts of question can be asked about genetic engineering (scientific, technological, moral, religious, political, economic, etc.) and that these questions have to be tackled in distinctively different ways (e.g. moral and religious questions cannot be settled by laboratory experiments or by statistical surveys).

The distinction between educating and informing has implications for the issue of labelling genetically engineered products, which has already been discussed in Part 2. Self-evidently, a label cannot possibly educate anyone; all that it can do is to offer a strictly limited amount of information about a particular product, but this can hardly count even as a 'public information' exercise. To put the words 'product of gene technology', for example, on a packet of cheese is 'informative' only if it is read and understood by the person who buys it. The function of such labels, then, is perhaps best seen not as educational or even informative, but rather as signalling a possible choice for the consumer. The shopper who knows something of what 'gene technology' means and the issues it raises is enabled by means of appropriate labelling to exercise choice in selecting or rejecting the labelled product.

This still begs the question of how the knowledge that is a prerequisite of such choices is to be acquired. Producers and retailers cannot be expected to launch full-scale educational programmes about genetic engineering, but one promising trend is towards the provision of informative leaflets by retailers that could be useful in explaining to those consumers who wish to know more about the subject what the phrase on the label exactly means.

Education or persuasion?

The provision of information must play a part in any broader educational initiative, though, as we have just seen, it cannot constitute the whole of it. But whether we are talking about educating or informing the public about genetic engineering, what objectives or end-results are we envisaging?

Quite often those who advocate increasing public awareness and understanding seem to have a preconceived assumption that more 'education' in this area will (or should) inevitably lead the recipients of this 'education' to a particular set of conclusions. This is well illustrated by some of the positions that were outlined earlier in this chapter. For the 'conspiracy theorists', for example, the end-result will involve the public coming to see through the machinations of politicians and big business and consequently to oppose or reject the new technology. For the 'optimists', more 'education' will (or should) lead people to see the benefits of genetic engineering and so accept it.

Yet neither of these opposed viewpoints is strictly speaking concerned with *education* as such, for education is by definition open-ended and unpredictable. You cannot educate anybody *to become* a Marxist or a monetarist or an anti-vivisectionist or an atheist, because the end-product of education cannot be specified and guaranteed in that way. Certainly there are methods of trying to get people to hold certain views and values and to behave in particular ways; these might include persuasion, coercion, suggestion, propaganda, advertising, indoctrination or conditioning. But these are not methods of *education.*[6] To set out with the aim of producing a particular set of beliefs and attitudes about a specific issue rarely counts as education, and those who wish to achieve such an end cannot claim to be advocating education. This is not merely a trivial, linguistic point. Two different activities need to be distinguished here (however we label them): on the one hand providing people with knowledge and understanding for them to use as tools in making informed decisions, and on the other persuading them to adopt a particular stance.

Because genetic engineering is such an emotive subject, it tends to attract persuasive rather than educational approaches, and such persuasion often uses emotive language, the function of which is to exhort rather than inform. If genetic engineering is described, for example, as 'tampering with the stuff of life', this description is clearly intended to arouse a negative response.

Another important context in which the distinction between education and persuasion must be drawn and the role of emotive language recognised is the treatment of genetic engineering by the media. Newspapers, television and radio programmes provide an obvious and influential medium for educating and informing the public, and if the adult population is to gain more understanding of genetic engineering it is most likely to do so from such sources. The subject is gaining an increasingly extensive coverage in all areas of the media, and as might be expected this coverage varies enormously in level and quality. While serious and informative articles and programmes can at times be found dealing with the scientific and non-scientific aspects in a reasonably balanced way, genetic engineering unfortunately lends itself to over-sensational treatment and over-dramatic presentation, often highlighting its most threatening and sinister features. Emotive language is again frequently used here, intended to promote a particular 'line' or 'angle' on the subject being dealt with. This technique produces an exaggerated impression of both the potential risks and the benefits of genetic engineering (e.g. by using headlines about 'mutant mutton' on the one hand or about 'miracle cancer gene break-through' on the other). Such approaches cannot claim to be 'educating' the public.

How can people be educated or informed about genetic engineering?

Methods of persuasion seem to be more commonly used and more easily devised than methods of education in this area. Genetic engineering does not, however, stand alone in this respect, for any subject that is controversial and raises a variety of moral, religious and social issues poses difficult problems for education; other obvious examples include abortion, euthanasia, sexual behaviour, animal rights, the disciplining of children and the punishment of criminals. Different views can be taken and defended on these issues about what is the right and wrong course of action, but no conclusive proof can be provided to establish what really and definitively *is* right and wrong. We can argue about these matters of right and wrong, but we cannot deduce from the 'facts' of the case the correct answer about what ought or ought not to be done about these 'facts'.

So what can education hope to achieve in such cases? People can be taught (in schools and elsewhere) the facts about genetic engineering, but this factual information will not give automatic answers to the moral, religious and social questions that the technology is generating. But if it is those questions that are prompting the demand for greater public awareness, what sort of education is possible here? How, in other words, can education tackle such controversial issues?

If the aim of such an educational programme is to enable people to make their own informed judgments and decisions about genetic engineering, clearly a balanced and unbiased approach will be needed that does not express support for one particular viewpoint. But is such an approach possible? Can a neutral stance be adopted by the educator?

This question has been closely examined by teachers and educationalists in recent years and has stimulated a lively debate. One particular teaching project for schools,[7] for example, envisaged the role of the teacher in dealing with controversial issues as that of a neutral chairperson who has 'responsibility for quality and standards in learning' and who presents relevant material to the group, but who takes no substantive part in the discussion and deliberately avoids promoting his or her own view. In addition, the teacher is not allowed to show any form of verbal or non-verbal approval or disapproval of any opinions expressed, nor to push the group towards any form of consensus. However, critics of this approach have claimed that such neutrality is neither possible nor desirable;[8] it is impossible because the selection and presentation of 'relevant' material cannot avoid implicit value-judgments, and it is undesirable because educators should not try to operate in a totally value-free, uncommitted ethos.

A further objection is that this approach may be taken to imply that any opinion on a controversial issue is as sound and valid as any other, thus reflecting the currently fashionable 'subjectivist' view that 'what I think is right is right for me' and that moral beliefs are consequently merely matters of personal taste. When pushed to its extreme, however, few would be willing to accept the implications of this view. If a scientist thought, for example, that it was 'right for him' to develop a form of biological warfare that singled out women and young children for a particularly excruciating death, would we be happy to shrug our shoulders and say that our moral repugnance at this project was 'just our opinion' or 'a matter of taste'? There are certain generally agreed presuppositions about ethical decisions and actions, among the most important of which are that they require us to consider other people's interests and to put ourselves in the place of those most affected.[9] Ethical judgments and decisions are also usually thought to possess other general characteristics of a rational and non-arbitrary kind. We are not, for example, making an *ethical* judgment about engineering transgenic animals if we make a decision simply on the basis of how we happen to be feeling that day, or of how fond we are of furry animals (excepting rats and mice), or of whether the coin we toss comes down heads or tails.[10]

Increased awareness and understanding of genetic engineering and the controversial issues it raises, therefore, will require not only some basic knowledge of the technology itself but also some appreciation of how to go about the business of making ethical decisions about complex questions. Such decisions need to be seen neither as mere matters of personal taste where all views are equally sound, nor as conclusively correct conclusions that cannot be queried or debated. Perhaps the best parallel is with *legal* decisions: one can never prove conclusively that the accused is innocent or guilty, but a balanced, non-arbitrary judgment has to be made on the basis of the evidence and in accordance with legal rules and principles which try to ensure that personal opinions, tastes and prejudices do not determine the outcome.

So educating people about genetic engineering cannot be reduced to simply 'telling them the facts'. Making decisions and judgments *about* those facts is just as difficult and demanding as acquiring the factual information (and probably more so). It is not a skill that just comes naturally; like any other skill it has to be demonstrated, learned and practised. So how can people be encouraged to develop it?

Again we need to distinguish here between educational activities that are school or college based and those that are not. If we are considering the education of students in schools and colleges, then there is at least a clear institutional context in which to tackle the problem. What must be guarded against is the temptation to label genetic engineering as a purely 'scientific' item on the curriculum and to ignore the many other dimensions that have been illustrated in this book.

When and how these other dimensions should be dealt with is not, however, self-evidently obvious. It would be possible, for example, for the scientific aspects to be taught by specialist science teachers, and for the non-scientific, controversial areas to be handled by teachers whose responsibility was for moral and religious education. Or a more integrated, topic-based approach could be used, where the scientific and non-scientific questions were dealt with side by side. Whichever approach is adopted, it needs to be appreciated that teaching young people how to tackle controversial issues requires skilled, specialist teachers, experienced in a wide range of methods and approaches that avoid the extremes of an 'anything-goes' subjectivism on the one hand and of a prescriptive authoritarianism on the other.

A number of specific suggestions have been made.[11] One possibility is that teachers should present to their students as many sides of the controversy as possible without, at least initially, indicating which they personally support. One difficulty, though, is that, by its very nature, a controversial issue, such as whether genetic engineering in a particular instance is acceptable or not, is one for which a teacher may find it particularly difficult to give a balanced presentation. Indeed, precisely what would constitute a balanced presentation may itself be open to debate. A further difficulty with this approach is that the lesson may end up being very didactic and fail to engage the interest and involvement of many in the class. A second possibility is that the teacher acts as a facilitator. Different points of view are elicited from the students and backed up by resource material.

Whatever approach is used, teachers need to keep in mind that the aim is not only to increase knowledge and understanding about genetic engineering but also to help students to develop the skills that will allow them to clarify the issues for themselves and assess the rightness of particular courses of action.

The educational problems for schools here are considerable but by no means insurmountable; they are, in principle, no different from those arising from the teaching of any scientific subject with wider controversial implications. Much more difficult is the task of attempting this kind of education outside the institutional context for adults who are not at school or college. Mention has already been made of the role of the media in this connection and the communication of information to consumers, and these two channels can make a significant and influential contribution in raising public awareness. Another more innovative method has also been tried, however, both in the UK and elsewhere, and we will conclude this brief review of the educational problems by describing and commenting upon it.

In November 1994, the UK had its first experience of a consensus conference. Consensus conferences started in Denmark in the 1980s and have since spread elsewhere. Along with 'citizen's juries', 'deliberative opinion polls' and a number of other approaches, they are experiments in democracy.[12] In essence a consensus conference involves getting together a lay panel that then produces a report on a particular topic of current importance. In the case of the first UK consensus conference, the topic was plant biotechnology. Benefiting from funding provided by the Biotechnology and Biological Sciences Research Council, and under the chairpersonship of Professor John Durant, Assistant Director of the London Science Museum, a panel of 16 lay members was assembled.

The lay members were chosen from over 300 people who had replied to advertisements placed in regional newspapers and relayed on various radio stations. Two criteria were used in the selection of panel members: first, each of them should not have any existing connections with biotechnology companies or environmental pressure groups; secondly, collectively they should constitute a panel balanced with respect to gender, age, education, occupation, ethnicity, etc.

After two briefing meetings, the panel identified seven key questions:

- What are the key benefits and/or risks of modern plant biotechnology?
- What possible impact could plant biotechnology have on the consumer?
- What possible impact could plant biotechnology have on the environment?
- What moral problems are raised by plant biotechnology?
- Why are patenting and intellectual property rights such a feature of plant biotechnology?
- How can we ensure that plant biotechnology benefits rather than harms the developing world now and in the future?
- What are the prospects for effective regulation of plant biotechnology?

To help them answer these questions, the lay panel called on expert witnesses. These witnesses were summoned to the consensus conference itself, which was held over two and a half days at a widely publicised central London venue. The general public were free to attend – and several hundred did so – but the proceedings themselves were largely under the control of the lay panel.

The climax of the consensus conference came on the morning of the third day when the lay panel published the report it had written the night before! This report, which runs to eight A4 pages, has been widely welcomed and praised by almost everyone who has read it. This is not the place either to comment in detail on the report itself, nor to suggest specific refinements to the way in which the process leading to its production was undertaken. What is worth emphasising is that experience in Denmark, the Netherlands and the UK suggests that the governments of different countries vary greatly in the significance they attach to the conclusions of particular consensus conferences. Our hope is that, whether via consensus conferences or other means, governments and other legislative bodies, such as the European Parliament, will take seriously the need to obtain the informed views of the general public before framing legislation in the area of genetic engineering.

Conclusions

Education in the field of genetic engineering is essential. Such education enables all of us to be better informed both about the science and about the ethics of genetic engineering. Education is not the same as information, though reliable information aids education. Education takes place both within educational establishments (notably schools, colleges and universities) and more widely – through such media as newspapers, journals, books, television and radio.

It is of the nature of education that it is open-ended: its results cannot precisely be predicted or determined in advance. Nor is education merely about acquiring 'facts', even when what the facts are can be agreed upon. To become more educated is, in a very real sense, to become more human. For this reason, it is incumbent on each of us to educate ourselves, and on those in authority over us, such as national governments, to enable us to be educated. In the case of genetic engineering, the pace of technological change makes these needs particularly pressing.

We make no excuses for focusing our final conclusions of this book upon the need for education and communication. More specific conclusions about particular applications of genetic engineering have been presented in Part 2, but a major argument of this book has been that it is over-simplistic to attempt to reach any overall conclusion about the rightness or wrongness of genetic engineering *per se*. Instead we have suggested and demonstrated a case-by-case approach, which enables particular applications to be judged on their merits. The methods of judgment to be used, however, are far from clearcut, and remain open to further discussion, debate and refinement. This process is one in which the public must be encouraged and enabled to participate in on an informed basis if rational and responsible decisions are to be made about developments that will significantly affect the lives of each one of us.

Notes

1 INTRODUCTION

1. The terms 'modern biotechnology' and 'genetic modification' are probably used, on at least some occasions, in an attempt to sound 'softer' than the other terms and so gain public acceptance. Strictly speaking, the term 'recombinant DNA technology' is narrower than 'genetic engineering'. For example, antisense technology (discussed on p. 134) need not involve recombinant DNA technology but is manifestly an example of genetic engineering. At the time of writing, no internationally accepted definition of genetic engineering exists and most national and international legislation on the subject includes its own definition. In reality, though, there is almost never any controversy as to whether something is an example of genetic engineering or not. As a rule of thumb, any change to the genetic make-up of an organism resulting from the direct insertion of genetic material either from another organism or constructed in the laboratory is an instance of genetic engineering.
2. Other definitions exist but are not very different. The scope of biotechnology is addressed by Bains, W. (1993). *Biotechnology from A to Z*. Oxford: Oxford University Press.
3. The information on the early history of traditional biotechnology is taken from Reiss, M. J. (1993). *Science Education for a Pluralist Society*. Milton Keynes: Open University Press.
4. Redman, C. L. (1978). *The Rise of Civilization*. New York: W. H. Freeman.

5. We are grateful to Nicholas Postgate who kindly provided information about the dating and interpretation of this tablet.
6. Stich, S. P. (1978). The recombinant DNA debate. In *Philosophy & Public Affairs*, vol. 7(3). Reprinted in Ruse, M. (ed.) (1989). *Philosophy of Biology*, pp. 229–43. New York: Macmillan.

2 THE PRACTICALITIES OF GENETIC ENGINEERING

1. For further information on the structure of the genetic material, with copious illustrations, see the relevant chapters in any of the following textbooks: Campbell, N. A. (1990). *Biology*, 2nd edn. Redwood City, CA: Benjamin/Cummings; Purves, W. K., Orians, G. H. & Heller, H. C. (1992). *Life: The Science of Biology*, 3rd edn. Sunderland, MA: Sinauer; Roberts, M. B. V., Reiss, M. J. & Monger, G. (1993). *Biology: Principles and Processes*. Walton-on-Thames, UK: Nelson; Starr, C. & Taggart, R. (1989). *Biology: The Unity and Diversity of Life*, 5th edn. Belmont, CA: Wadsworth.
2. For further information on the techniques of genetic engineering, see Wymer, P. E. O. (1988). *Genetic Engineering*. Cambridge: Hobsons; Weaver, R. F. & Hedrick, P. W. (1992). *Genetics*, 2nd edn. Dubuque, IA: Wm. C. Brown; Bains, W. (1993). *Biotechnology from A to Z*. Oxford: Oxford University Press; Williams, J. G., Ceccarelli, A. & Spurr, N. (1993). *Genetic Engineering*. Oxford: BIOS Scientific.
3. Mitani, K. & Caskey, C. T. (1993). Delivering therapeutic genes – matching approach and application. *Trends in Biotechnology*, 11, 162–6.

3 MORAL AND ETHICAL CONCERNS

1. Talbot, C. (1991). Transgenic animals and the reduction of life, *Humane Education Newsletter*, 2(1), 12.
2. Cross, B. (1991). Look at it this way. *Outlook on Agriculture*, 20(2), 71–2.
3. Habgood, J. (1993). Ethical restraints on biotechnological inventiveness. In *Biotechnology – Friend or Foe?* Report of the

First Annual Meeting of the BioIndustry Association, Queen Anne's Gate, London.

4. Lee, T. R., Cody, C. & Plastow, E. (1985). *Consumer Attitudes towards Technological Innovations in Food Processing.* Guildford, UK: University of Surrey.

5. Hoban, T. J. & Kendall, P. A. (1992). *Consumer Attitudes about the use of Biotechnology in Agriculture and Food Production*, pp. 4–5. Raleigh, NC: North Carolina State University.

6. The philosopher Peter Singer, for example, begins his important book *Practical Ethics* (Cambridge University Press, 1979) by saying 'This book is about practical ethics, that is about the application of ethics or morality – I shall use the words interchangeably . . .'.

7. Some philosophers, however, use 'moral' to refer to questions of rights, duties and obligations, reserving 'ethical' for more general matters concerning 'the good life'.

8. Flew, A. (ed.) (1979). *A Dictionary of Philosophy*, pp. 104–5. London: Pan.

9. Reference 8, p. 105.

10. Stich, S. P. (1989). The recombinant DNA debate. In *Philosophy of Biology*, ed. M. Ruse, pp. 229–30. New York: Macmillan.

11. Rifkin, J. (1987). Biotechnology: major societal concerns. In *Public Perceptions of Biotechnology*, ed. L. R. Batra & W. Klassen, p. 60. Bethesda, MD: Maryland Agricultural Research Institute.

12. Rifkin, J. (1985). *Declaration of a Heretic*, p. 48. London: Routledge & Kegan Paul.

13. Summary of the report by a Swedish government committee (SOU) (1992). *Genetic Engineering – A Challenge*, p. 11. Göteborg, Sweden: Graphic Systems AB.

14. Krimsky, S. (1982). *Genetic Alchemy.* Cambridge, MA: MIT Press.

15. Quoted in Reference 14, p. 322.

16. Reference 12, p. 71.

17. Reference 10, pp. 231–2.

18. Godown, R. D. (1987). The science of biotechnology. In *Public Perceptions of Biotechnology*, ed. L. R. Batra &

W. Klassen, pp. 23–4. Bethesda, MD: Maryland Agricultural Research Institute.

19. Reference 14, p. 77.
20. Reference 14, p. 285.
21. Reference 14, p. 139.
22. Fox, M. (1990). Transgenic animals: ethical and animal welfare concerns. In *The Bio-Revolution: Cornucopia or Pandora's Box*, ed. P. Wheale & R. McNally, pp. 31–45. London: Pluto Press.
23. Ridley, M. (1985). *The Problems of Evolution*. Oxford: Oxford University Press.
24. Cupitt, D. (1975). Natural evil. In *Man and Nature*, ed. H. Montefiore, p. 100. London: Collins.
25. Reference 12, p. 53.
26. Thompson, J. L. (1983). Preservation of wilderness and the good life. In *Environmental Philosophy*, ed. R. Elliott & A. Gare, p. 89. Milton Keynes: Open University Press.
27. World Council of Churches (1988). *Integrity of Creation*, p. 3. Geneva: World Council of Churches.
28. Kant, I. (1909). *Fundamental Principles of the Metaphysic of Morals*, 6th edn (trans. Abbott, K. T.), p. 47. London: Longman.
29. Holland, A. (1990). The biotic community: a philosophical critique of genetic engineering. In *The Bio-Revolution: Cornucopia or Pandora's Box?*, ed. P. Wheale & R. McNally, p 170. London: Pluto Press. This paper offers a very useful and thoughtful discussion of a number of issues.
30. Some of the material in this chapter has been drawn from Straughan, R. (1995). Ethical aspects of crop biotechnology. In *Issues in Agricultural Bioethics*, ed. T. B. Mepham, G. A. Tucker & J. Wiseman. Nottingham, UK: Nottingham University Press.

4 THEOLOGICAL CONCERNS

1. Jones, I. H. (1991). Theology and genetics. Paper given to a Consultation on Moral and Theological Questions in Genetic Manipulation at Luton Industrial College, p. 8.

2. Rifkin, J. (1993). Foreword to Kimbrell, A. (1993). *The Human Body Shop: The Engineering and Marketing of Life*, p. x. London: HarperCollins*Religious*.
3. For example, Independent Television Authority (1970). *Religion in Britain and Northern Ireland*. London: ITA; Greeley, A. M. (1974). *Ecstasy: A Way of Knowing*. Englewood Cliffs, NJ: Prentice-Hall; Argyle, M. & Beit-Hallahmi, B. (1975). *The Social Psychology of Religion*. London: Routledge and Kegan Paul; Hay, D. (1982). *Exploring Inner Space: Scientists and Religious Experience*. Harmondsworth: Penguin; Watts, F. & Williams, M. (1988). *The Psychology of Religious Knowing*. Cambridge: Cambridge University Press; Brierley, P. (1991). *Christian England*. London: MarcEurope; Kerkhofs, J. (1993). Will the Churches meet the Europeans? *The Tablet*, 18 September, 1184–7. These surveys are mostly carried out in industrialised democracies, but there can be little doubt that the results would be similar, or show even a higher proportion of the population to have a religious faith, in other countries.
4. A 1992 study of 1228 telephone interviews in the USA showed that the stronger a person's religion was to them, the more likely they were to believe that genetic engineering was wrong (Hoban, T. J. & Kendall, P. A. (1992). *Consumer Attitudes about the Use of Biotechnology in Agriculture and Food production*. Raleigh, NC: North Carolina State University). The same survey showed that while the most common reason for opposing biotechnology involved concerns that it would 'threaten the balance of nature', a significant proportion mentioned that biotechnology was 'not natural', somehow 'against God's will' or 'contrary to [their] religious beliefs'.

It is difficult to quantify the importance of religious beliefs to such attitudes, but similar studies in New Zealand, Japan and Europe also show that religious beliefs are among the most widely cited reason why some people consider genetic engineering to be unacceptable (Couchman, P. K. & Fink-Jensen, K. (1990). *Public Attitudes to Genetic Engineering in New Zealand*. Christ Church, New Zealand: Department of Scientific and Industrial Research, Crop Research Division; Dixon, B. (1991). Biotech a plus according to European poll.

Bio/Technology, 9, 16; Macer, D. R. J. (1992). Public acceptance of human gene therapy and perceptions of human genetic manipulation. *Human Gene Therapy*, 3, 511–8).

For other data on people's attitudes to genetic engineering, see Yearley, S. (n.d.). *Biotechnology: For or Against. Public Beliefs, Attitudes and Understanding.* A Report of the Workshop convened by the ESRC for the Biotechnology Joint Advisory Board. Swindon: Economic & Social Research Council; Rothenberg, L. (1994). Biotechnology's issue of public credibility. *Trends in Biotechnology*, 12, 435–8; Brown, C. M. (1994). *Consumer Attitudes to Biotechnology in Agriculture and Food: A Critical Review.* Watford: Institute of Grocery Distribution.

5. No attempt will be made here to define exactly what is meant by a religion. Some would argue that even atheism or Marxism are religions. We use the word in its everyday sense. Most religious adherents have a sense of the transcendent, frequently of a God (or gods) who exists and has some connection with, or relationship to, our everyday lives. By 'theological concerns' we mean those questions and issues that arise from holding religious beliefs.

6. For a more detailed discussion, see Berg, J. (1993). How could ethics depend on religion? In *A Companion to Ethics*, ed. P. Singer, pp. 525–33. Oxford: Blackwell.

7. The headings used here are taken from Smart, N. (1989). *The World's Religions: Old Traditions and Modern Transformations.* Cambridge: Cambridge University Press. (An early attempt to delineate the essential characteristics shared by religions was made by King, M. B. & Hunt, R. A. (1972). Measuring the religious variable: reflections. *Journal for the Scientific Study of Religion*, 11, 240–51.) Some of the information about different religions is based on Smart and some on Hinnells, J. R. (1991). *A Handbook of Living Religions.* London: Penguin.

8. The principle of nonmaleficence is discussed by Beauchamp, T. L. & Childress, J. F. (1989). *Principles of Biomedical Ethics*, 3rd edn. Oxford: Oxford University Press. Utilitarianism and other approaches are discussed in Singer, P. (ed.) (1993). *A Companion to Ethics.* Oxford: Blackwell.

9. The relationship between secular and religious ethics is discussed from different viewpoints by Smart, N. (1958). *Reasons and Faiths: An Investigation of Religious Discourse, Christian and Non-Christian.* London: Routledge & Kegan Paul; Mackie, J. L. (1977). *Ethics: Inventing Right and Wrong.* Harmondsworth, UK: Penguin; Cupitt, D. (1988). *The New Christian Ethics.* London: SCM Press; Nielsen, K. (1990). *Ethics Without God.* Buffalo, New York: Prometheus Books; Gill, R. (1991). *Christian Ethics in Secular Worlds.* Edinburgh: T&T Clark.

10. For a fuller discussion of natural law see Finnis, J. (1980). *Natural Law and Natural Rights.* Oxford: Clarendon Press; Brooke, J. H. (1991). *Science and Religion: Some Historical Perspectives.* Cambridge: Cambridge University Press; Brooke, J. H. (1992). Natural law in the natural sciences: the origins of modern atheism? *Science and Christian Belief,* 4, 83–103.

11. Brower, D. (Friends of the Earth) (1987). Interview with *San Francisco Chronicle,* quoted in Jukes, T. H. (1988). Hazards of biotechnology: facts and fancy. *Journal of Chemical Technology and Biotechnology,* 43, 249.

12. Genesis 1.28.

13. Genesis 1.27.

14 White, L. (1967). The historical roots of our ecologic crisis. *Science,* 155, 1203–7.

15. Revelation 21.1.

16. Prime, R. (1992). *Hinduism and Ecology: Seeds of Truth.* London: Cassell. See also Chapple, C. (1994). Contemporary Jaina and Hindu responses to the ecological crisis. In *An Ecology of the Spirit: Religious Reflection and Environmental Consciousnes,* ed. M. Barnes, pp. 209–18. Lanham: University Press of America.

17. Prime, R. Reference 16, p. 32.

18. Leviticus 25.4.

19. Leviticus 25.23.

20. Job 39.5–10.

21. Rose, A. (1992). *Judaism and Ecology,* p. 27. London: Cassell.

22. Batchelor, M. & Brown, K. (ed.) (1992). *Buddhism and Ecology.* London: Cassell. See also Grosnick, W. (1994). The

Buddahood of the grasses and the trees: ecological sensitivity or scriptural misunderstanding? In *An Ecology of the Spirit: Religious Reflection and Environmental Consciousnes*, ed. M. Barnes, pp. 197–208. Lanham: University Press of America.

23. Tanahashi, K. (ed.) (1988). *Dogon. Moon in a Dewdrop*, pp. 106–7. Shaftesbury: Element Books.
24. Thich Nhat Hanh (1988). *The Sun My Heart*, p. 90. Berkeley: Parallax Press.
25. Innes, K. (1987). *Caring for the Earth – The Environment, Christians and the Church*. Bramcote, Nottingham: Grove Books; Bradley, I. (1990). *God is Green: Christianity and the Environment*. London: Darton, Longman and Todd; World Council of Churches (1990). *Giver of Life – Sustain Your Creation. Report of the Pre-Assembly Consultation on Sub-theme 1*, Kuala Lumpur, Malaysia. Reprinted in *The Ecumenical Review*, July–October 1990; DeWitt, C. B. (1994). *Earth-Wise: A Biblical Response to Environmental Issues*. Grand Rapids, MI: CRC Publications.
26. Qur'an 2.204–6; 24.41; 30.30; Khalid, F. & O'Brien, J. (ed.) (1992). *Islam and Ecology*. London: Cassell.
27. Hinnells, J. R., see Reference 7; Smart, N., see Reference 7.
28. Peacocke, A. R. (1979). *Creation and the World of Science: The Bampton Lectures, 1978*. Oxford: Clarendon Press.
29. Moltmann, J. (1985). *God in Creation: An Ecological Doctrine of Creation: the Gifford Lectures 1984–1985*. London: SCM Press; Barbour, I. G. (1990). *Religion in an Age of Science: The Gifford Lectures 1989–1991*, vol. 1. London: SCM Press. Peacocke (Reference 28) further explores other understandings of humanity's role in creation including those of priest, prophet, fellow-sufferer and lover. The more general question of God's relationship to the universe is considered by process theologians and feminist theologians, *inter alia*. See Cobb, J. B. & Griffin, D. R. (1969). *Process Theology: An Introductory Exposition*. Belfast: Christian Journals; Pittinger (n.d.). What Does God Do? An Introduction to Process Theology. *Modern Churchmen's Union Pamphlet*, 17; McFague, S. (1993). *The Body of God: An Ecological Theology*. London: SCM Press.

30. Reference 28, p. 305.
31. Letter from Dr Donald M. Bruce to Michael J. Reiss, 5 December 1994.
32. Nelson, J. R. (1994). *On the New Frontiers of Genetics and Religion*. Grand Rapids, MI: William B. Eerdmans. See also Dyson, A. (1994). Genetic engineering in theology and theological ethics. In *Ethics and Biotechnology*, ed. A. Dyson & J. Harris, pp. 259–71. London: Routledge.
33. Linzey, A. (1990). Human and animal slavery: a theological critique of genetic engineering. In *The Bio-Revolution: Cornucopia or Pandora's Box*, ed. P. Wheale & R. McNally, pp. 175–88. London: Pluto Press (see p. 180). The same point is made in Linzey, A. (1994). *Animal Theology*. London: SCM. See also Montefiore, H. (1993). Rights in animals. *Church Times*, 19 February.
34. Letter from K. S. Satagopan to Michael J. Reiss dated 7 October 1993.
35. Heidegger, M. (1977). *The Question Concerning Technology and Other Essays. Translated and with an Introduction by William Lovitt*. New York: Harper Colophon.
36. Macer, D. (1990). Genetic engineering in 1990. *Science & Christian Belief*, 1, 25–40.
37. Berry, A. C. (1992). Genetic engineering and medical treatment. In *Genetic Engineering: Christian Responsibilities in God's World*, a conference organised by Christians in Science and the Christian Medical Fellowship, Paper 2, pp. 5–6. Transcript available from Christian Medical Fellowship, 157 Waterloo Road, London SE1 8XN.
38. Challis, P. (1992). *Genetic Engineering and its Applications: Some Theological and Ethical Reflections*, pp. 39–40. Cambridge: Wesley House. Celia Deane-Drummond (in Deane-Drummond, C. (1995). Reshaping our environment: implications of the new biotechnology. *Theology in Green*, 5, 19–33) argues that, 'A theological approach encourages those involved to see the wider social and religious consequences of these decisions. It does not necessarily ban all genetic engineering, but seeks to transform it so that it more clearly represents a fully human enterprise' (p. 26).

39. Cole-Turner, R. (1993). *The New Genesis: Theology and the Genetic Revolution*. Louisville, KY: Westminster/John Knox Press.
40. See, for example, Williams, B. (1992). Must a concern for the environment be centred on human beings? In *Ethics and the Environment*, ed. C. C. W. Taylor, pp. 60–8. Oxford: Corpus Christi College.

5 THE GENETIC ENGINEERING OF MICROORGANISMS

1. Deakin, R. (1990). BST: The first commercial product for agriculture from biotechnology. In *The Bio-Revolution: Cornucopia or Pandora's Box*, ed. P. Wheale & R. McNally, p. 72. London: Pluto Press.
2. Spallone, P. (1992). *Generation Games: Genetic Engineering and the Future for Our Lives*, p. 23. London: The Women's Press.
3. Bibliographies of human insulin 1981–1988 (dated 3 September 1991) and 1989–1992 (dated 24 March 1994) obtained from British Diabetic Association, Information Science Department, 10 Queen Anne Street, London W1M 0BD. See also Pickup, J. (1989). Human insulin: problems with hypoglycaemia in a few patients. *British Medical Journal*, 299, 991–3; Wolff, S. P. (1992). Trying times for human insulin. *Nature*, 356, 375–6.
4. British Diabetic Association (1991). *People with Diabetes and Changes in Hypo Warnings*. London: British Diabetic Association.
5. Bristow, A. F. (1993). Recombinant-DNA-derived insulin analogues as potentially useful therapeutic agents. *Trends in Biotechnology*, 11, 301–5 (particularly p. 302).
6. Day, S. (1993). Jelly capsules offer end to diabetics' daily dose. *New Scientist*, 4 September, 16.
7. Tanner, J. M., Rattray Taylor, G. *et al.* (1968). *Growth*. Netherlands: Time-Life International.
8. Kimbrell, A. (1993). *The Human Body Shop: The Engineering and Marketing of Life*. London: HarperCollins *Religious*; Horizon (1995). *Too Big Too Soon?* London: British Broadcasting Corporation.

9. Adult Patient Consent Form, Study Number 91–CH–46, p. 5, paragraph 9. Bethesda, MD: The National Institutes of Health. Cited by Kimbrell, A., Reference 8, p. 146.

10. Lippman, A. (1993). Jurassic Park opens gates to public discussion. *GeneWATCH*, 9(1–2), 4–5 (particularly p. 5).

11. Werth, B. (1991). How short is too short, marketing human growth hormone. *New York Times magazine*, 16 June, 14. Cited by Kimbrell, A., Reference 8, p. 154.

12. BST is also known as BGH (bovine growth hormone). A detailed review of the arguments for and against the use of genetically engineered BST is provided by Wheale, P. & McNally, R. (ed.) (1990). *The Bio-Revolution: Cornucopia or Pandora's Box*. London: Pluto Press.

13. Monsanto (1986). *Bovine Somatotropin: An Introductory Guide to a New Farming Tool*. Brussels: Monsanto Europe SA/NV.

14. Monsanto (1993). *Posilac (Sterile Sometribove Zinc Suspension)*. St Louis: Monsanto. See also Coghlan, A. (1994). Milk hormone data bottled up for years. *New Scientist*, 22 October, 4.

15. Reference 13, p. 9.

16. Reference 13, p. 2.

17. The suggestion that BST might trigger premature growth in infants, breast development in children and breast cancer in women comes from Epstein, S. E., Professor of Environmental & Occupational Medicine, University of Illinois. Cited by Compassion in World Farming (n.d.) [1994] *Fact Sheet: BST*. Petersfield, UK: Compassion in World Farming. Mepham's arguments can be found in Mepham, T. B. (1992). Public health implications of bovine somatotrophin use in dairying: discussion paper. *Journal of the Royal Society of Medicine*, 85, 736–9; Mepham, T. B., Schofield, P. N., Zumkeller, W. & Cotterill, A. M. (1994). Safety of milk from cows treated with bovine somatotrophin. *Lancet*, 344, 197–8.

18. Coghlan, A. (1994). Keep milk hormone ban, say farmers. *New Scientist*, 13 August, 8.

19. Kleiner, K. (1994). US bans 'hormone free' milk label. *New Scientist*, 26 February, 5.

20. Kleiner, K. (1994). Milk hormone dispute boils over into court. *New Scientist*, 30 April, 6–7.
21. Bishop, D. H. L., Cory, J. S. & Possee, R. D. (1992). The use of genetically engineered virus insecticides to control insect pests. In *Release of Genetically Engineered and other Micro-organisms*, ed. J. C. Fry & M. J. Day, pp. 137–46. Cambridge: Cambridge University Press; also based on Cory, J. S. & Bishop, D. H. L. (1995). Biopesticides. In *Issues in Agricultural Bioethics*, ed. T. B. Mepham, G. A. Tucker & J. Wiseman, pp. 135–49. Nottingham, UK: Nottingham University Press.
22. Coghlan, A. (1994). Will the scorpion gene run wild? *New Scientist*, 25 June, 14–15; Coghlan, A. (1994). Scorpion gene virus on trial in Oxford. *New Scientist*, 3 December, 11.
23. Coghlan, A. (1994). Will the scorpion gene run wild? *New Scientist*, 25 June, 15.
24. Payne, C. C. & Lynch, J. M. (1988). Biological control. In *Micro-organisms in Action: Concepts and Applications in Microbial Ecology*, 2nd edn, ed. J. M. Lynch & J. E. Hobbie, pp. 261–87. Oxford: Blackwell Scientific.
25. Lindow, S. E. (1985). Ecology of *Pseudomonas syringae* relevant to the field use of ice-deletion mutants constructed *in vitro* for plant frost control. In *Engineered Organisms in the Environment: Scientific Issues*, ed. H. O. Halvorson, D. Pramer & M. Rogul, pp. 23–35. Washington, DC: American Society for Microbiology.
26. Krimsky, S. (1991). *Biotechnics and Society: The Rise of Industrial Genetics*. New York: Praeger.
27. Council for Responsible Genetics, 5 Upland Road, Suite 3, Cambridge, MA 02140, USA.
28. Fincham, J. R. S. & Ravetz, J. R. (1991). *Genetically Engineered Organisms: Benefits and Risks*. Milton Keynes: Open University Press.
29. Reference 28, p. 40.
30. Lindow, S. E. (1983). Methods of preventing frost injury caused by epiphytic ice-nucleation-active bacteria. *Plant Disease*, 67, 327–33 (particularly p. 332). Cited in Reference 26, p. 118.

31. Catroux, G. & Amarger, N. (1992). Rhizobia as soil inoculants in agriculture. In *Release of Genetically Engineered and other Micro-organisms*, ed. J. C. Fry & M. J. Day, pp. 1–13. Cambridge: Cambridge University Press.

32. Fry, J. C. & Day, M. J. (ed.) (1992). *Release of Genetically Engineered and other Micro-organisms*. Cambridge: Cambridge University Press; Wellington, E. M. H. & van Elsas, J. D. (ed.) (1992). *Genetic Interactions among Microorganisms in the Natural Environment*. Oxford: Pergamon Press; Edwards, C. (ed.) (1993). *Monitoring Genetically Manipulated Microorganisms in the Environment*. New York: Wiley; Dowling, D. N. & O'Gara, F. (1994). Metabolites of *Pseudomonas* involved in the biocontrol of plant disease. *Trends in Biotechnology*, 12, 133–41; Prosser, J. (1994). Tracking genetically engineered microorganisms. *Biological Sciences Review*, 6(4), 30–2. One very well-known ecologist who believed it possible that the widespread use of ice-minus bacteria might have very serious environmental consequences was Eugene P. Odum (see Odum, E. P. (1985). Biotechnology and the biosphere. *Science*, 229, 1338). For a response to Odum, see Lindemann, J., Warren, G. J. & Suslow, T. V. (1985). Ice-nucleating bacteria. *Science*, 231, 536. For ways of disabling genetically engineered microorganisms to reduce the chance, once they have done their job, of their persisting and spreading in the environment, see Ramos, J. L., Andersson, P., Jensen, L. B., Ramos, C., Ronchel, M. C., Díaz, E, Timmis, K. N. & Molin, S. (1995). Suicide microbes on the loose. *Bio/Technology*, 13, 35–7.

33. Commission of the European Communities (1994). Biotechnology and the White Paper on growth, competitiveness and employment – preparing the next stage. *European Biotechnology Information Service*, 4(2), 2–11; Coghlan, A. (1994). US loosens laws for testing modified pest killers. *New Scientist*, 23 July, 10; MacKenzie, D. (1994). Europe pushes for a genetically modified future. *New Scientist*, 11 June, 6.

34. Coghlan, A. Reference 33, p. 10.

35. MacKenzie, D. (1994) Mutant bacteria may escape from the mail. *New Scientist*, 4 April, 6.

36. Witt, S. (1993) Keep your eye on the tortoise. Transcript of a talk given at the Conference *Symbol, Substance and Science: The Societal Issues of Food Biotechnology*, held at the North Carolina Biotechnology Center, June 28–9, pp. 22–30.

37. CWS Co-op Brand, Customer Relations, P.O. Box 53, New Century House, Manchester M60 4ES, UK.

38. Letter from Andy Bond, Research Officer at the Vegetarian Society of the United Kingdom, to Michael J. Reiss, dated 19 November 1993.

39. Gould, S. J. (1982). Non-moral nature. *Natural History* (American Museum of Natural History), 91(2), 19.

40. Reference 39, p. 20

41. A delegate at the Ontario Federation of Agriculture (1988) is quoted as saying 'The cow is literally a milk machine and they don't last that long now' in Burton, J. L. & McBride, B. W. (1992). Recombinant bovine somatotropin: is there a limit for biotechnology in applied animal agriculture? *Journal of Agricultural and Environmental Ethics*, 2, 133

42. Rifkin, J. (1985) *Declaration of a Heretic*, p. 48. London: Routledge & Kegan Paul.

43. Clark, S. R. L. (1994). Modern errors, ancient virtues. In *Ethics and Biotechnology*, ed. A. Dyson & J. Harris, p. 23. London: Routledge.

44. See Straughan, R. (1989). *The Genetic Manipulation of Plants, Animals and Microbes*, pp. 25–6. London: National Consumer Council.

45. Quoted in Reference 41, pp. 154–5.

46. This question is debated in great detail in papers by Tweeten, L. and Comstock, G. (1991) in the *Journal of Agricultural and Environmental Ethics*, 4(2), 108–30.

47. Comstock, G. (1991). The costs and benefits of bGH may not be distributed fairly. *Journal of Agricultural and Environmental Ethics*, 4(2), 129.

48. Comstock, G. (1990). The case against bGH. In *Agricultural Bioethics: Implications of Agricultural Biotechnology*, ed. S. M. Gendel, A. D. Kline, D. M. Warren & F. Yates, p. 331. Ames, IA: Iowa State University Press.

49. Quoted in Taverne, D. (1990). *The Case for Biotechnology*, p. 38. London: Prima Europe.

50. The ethical issues concerning consumers' freedom of choice are discussed at greater length in Straughan, R. (1992). Freedom of choice: principles and practice. In *Your Food: Whose Choice?* ed. National Consumer Council, pp. 135–56. London: HMSO.

51. Report of the Committee on the Ethics of Genetic Modification and Food Use (1993). London: HMSO, p. 18.

6 THE GENETIC ENGINEERING OF PLANTS

1. Ken Barton, Vice-President of Research and Development at Agracetus talking about the prospects of genetically engineering plants to produce blue cotton. Quoted in Mestel, R. (1993). How blue genes could green the cotton industry. *New Scientist*, 31 July, 7.

2. Fax dated 31 October 1994 sent by Sue Mayer, Greenpeace to the lay panel at the first UK National Consensus Conference on Plant Biotechnology 2–4 November 1994, London. Sue Mayer was one of the invited experts who addressed the lay panel on 2 November 1994.

3. For reviews of recent techniques of genetic engineering in plants see Christou, P., Ford, T. L. & Kofron, M. (1992). The development of a variety-independent gene-transfer method for rice. *Trends in Biotechnology*, 10, 239–46; Bevan, M. W., Harrison, B. D. & Leaver, C. J. (ed.) (1993). The Production and Uses of Genetically Transformed Plants: Proceedings of a Royal Society Discussion Meeting held on 26 and 27 May 1993. *Philosophical Transactions of the Royal Society of London, Series B*, 342, 183–294; Maliga, P. (1993). Towards plastid transformation in flowering plants. *Trends in Biotechnology*, 11, 101–7; Gibson, S. & Somerville, C. (1993). Isolating plant genes. *Trends in Biotechnology*, 11, 306–13; Leemans, J. (1993). Ti to tomato, tomato to market. *Bio/Technology*, 11, S22–6; Yoder, J. I. & Goldsborough, A. P. (1994). Transformation systems for generating marker-free transgenic plants. *Bio/Technology*, 12, 263–7; Castillo, A. M.,

Vasil, V. & Vasil, I. K. (1994). Rapid production of fertile transgenic plants of rye (*Secale cereale* L.). *Bio/Technology*, 12, 1366–71.

4. In the laboratory, the plastic known as polyhydroxybutyrate has been produced by a genetically engineered plant species called *Arabidopsis thaliana* (thale cress). By the end of 1994, biologists at the Carnegie Institution of Washington in California had genetically engineered thale cress to produce up to 14% of its dry weight as plastic (Kleiner, K. (1995). Gene engineers grow better polymer plants. *New Scientist*, 14 January, 19; see also Poirier, Y., Nawrath, C. & Somerville, C. (1995). Production of polyhydroxyalkanoates, a family of biodegradable plastics and elastomers, in bacteria and plants. *Bio/Technology*, 13. 142–50). The significance of this work is that such plastics contain no chlorine and are biodegradable. They are, therefore, environmentally far more friendly than many of the plastics used to package materials. There is every likelihood that some plastics will soon be made from plant products, just as paper is.

5. Fromm, M. E., Stark, D. M., Austin, G. D. & Perlak, F. J. (1993). Improved agronomic and quality traits in transgenic crops: recent advances. *Philosophical Transactions of the Royal Society of London Series B*, 339, 233–7. See also related work to alter cytokinin (another plant growth regulator) production in tomatoes: Martineau, B., Summerfelt, K. R., Adams, D. F. & DeVerna, J. W. (1995). Production of high solids tomatoes through molecular modification of levels of the plant growth regulator cytokinin. *Bio/Technology*, 13, 150–4.

6. Zeneca Seeds (n.d.) [1994]. *The Tomato Story*. Haslemere, UK: Zeneca Seeds; Hall, L. N. *et al.* (1993). Antisense inhibition of pectin esterase gene expression in transgenic tomatoes. *Plant Journal*, 3(1), 121–9; Leemans, J. (1993). Reference 3.

7. Tucker, G. A. (1993). Improvement of tomato fruit quality and processing characteristics by genetic engineering. *Food Science and Technology Today*, 7(2), 103–8.

8. Department of Health Advisory Committee on Novel Foods and Processes (1991). *Report on Health and Social Subjects 38:*

Guidelines on the Assessment of Novel Foods and Processes.
London: HMSO.

9. Reference 8, p. 28.

10. Young, E. (1994). Altered tomato faces ban from British shops. *New Scientist*, 30 July, 9.

11. See, for example, Özcan, S., Firek, S. & Draper, J. (1993). Can elimination of the protein products of selectable marker genes in transgenic plants allay public anxieties? *Trends in Biotechnology*, 11, 219; Yoder, J. I. & Goldsborough, A. P. (1994). Transformation systems for generating marker-free transgenic plants. *Bio/Technology*, 12, 263–7; de Lorenzo, V. (1994). Designing microbial systems for gene expression in the field. *Trends in Biotechnology*, 12, 365–71.

12. Bartle, I. D. G. (1991). *Herbicide Tolerant Plants: Weed Control with the Environment in Mind.* Fernhurst, UK: Zeneca Seeds; Beard, J. (1993). Crops stand up to killer chemicals. *New Scientist*, 11 September, 16.

13. See, in particular, Comstock, G. (1989). Genetically engineered herbicide resistance, part one. *Journal of Agricultural Ethics*, 2, 263–306; Comstock, G. (1990). Genetically engineered herbicide resistance, part two. *Journal of Agricultural Ethics*, 3, 114–46.

14. Hartman, C. L., Lee, L., Day, P. R. & Tumer, N. E. (1994). Herbicide resistant turfgrass (*Agrostis palustris* Huds.) by biolistic transformation. *Bio/Technology*, 12, 919–23.

15. Kross, B. C., Burmeister, L. F. & Ogilvie, L. K. (1994). Mortality study among golf course superintendents. *Golf Course Management*, 62, 49–56.

16. For example, Dunwell, J. M. & Paul, E. M. (1990). Impact of genetically modified crops in agriculture. *Outlook on Agriculture*, 19(2), 103–9; Fax dated 1 November 1994 sent by Ed Dart, Zeneca Seeds, to the lay panel at the first UK National Consensus Conference on Plant Biotechnology 2–4 November 1994, London. Ed Dart was one of the invited experts who addressed the lay panel on 2 November 1994.

17. The prospects of such developments are discussed by John, M. E. & Stewart, J. McD. (1992). Genes for jeans: biotechnological advances in cotton. *Trends in Biotechnology*,

10, 165–70; Beck, C. I. & Ulrich, T. (1993). Biotechnology in the food industry. *Bio/Technology*, 11, 895–902; Kleiner, K. (1995). Hunt is on for supercrops. *New Scientist*, 11 February, 9.

18. Menon, S. (1994). Will DNA mark end of rice pest? *New Scientist*, 10 December, 15.
19. John, M. E. & Stewart, J. McD. (1992). Reference 17.
20. Coghlan, A. (1992). Toxic gene could rid maize of boring pest. *New Scientist*, 18 July, 21.
21. Horsch, R. B. (1993). Commercialization of genetically engineered crops. *Philosophical Transactions of the Royal Society of London Series B*, 342, 287–91.
22. Koziel, M. G., Beland, G. L., Bowman, C. *et al.* (1993). Field performance of elite transgenic maize plants expressing an insecticidal protein derived from *Bacillus thuringiensis. Bio/Technology*, 11, 194–200.
23. Coghlan, A. (1991). Gene transplants to zap sap-suckers. *New Scientist*, 14 December, 24.
24. Seneviratne, G. (1992). Gene transplant gives apricots a riper future. *New Scientist*, 14 March, 14.
25. Firoozababy, E., Moy, Y., Courtney-Gutterson, N. & Robinson, K. (1994). Regeneration of transgenic rose (*Rosa hybrida*) plants from embryogenic tissue. *Bio/Technology*, 12, 609–13; Wilson, L. (1994). No sex please, we're clones. *New Scientist*, 24/31 December, 26–8.
26. Beachy, R. (1993). Transferring genes. Part 1 of Session 4 of *Symbol, Substance and Science: The Societal Issues of Food Biotechnology*, held at the North Carolina Biotechnology Center, June 28–9, pp. 45–51.
27. Rissler, J. & Mellon, M. (1993). *Perils Amidst the Promise: Ecological Risks of Transgenic Crops in a Global Market.* Cambridge, MA: Union of Concerned Scientists.
28. Shorrocks, B. & Coates, D. (ed.) (1993). *British Ecological Society–Ecological Issues No. 4: The Release of Genetically-engineered Organisms.* Montford Bridge, Shrewsbury: Field Studies Council.
29. Holdgate, M. W. (1986). Summary and conclusions: characteristics and consequences of biological invasions.

Philosophical Transactions of the Royal Society Series B, 314, 733–42; Williamson, M. H. & Brown, K. C. (1986). The analysis and modelling of British invasions. *Philosophical Transactions of the Royal Society Series B*, 314, 505–22; Williamson, M. (1993). Invaders, weeds and the risks from genetically manipulated organisms. *Experientia*, 49, 219–24.

30. Williamson, M. (1994). Community response to transgenic plant release – predictions from British experience of invasive plants and feral crop plants. *Molecular Ecology*, 3, 75–9 (particularly p. 75).

31. Baker, H. G. (1965). Characteristics and modes of origins of weeds. In *The Genetics of Colonizing Species*, ed. H. G. Baker & G. L. Stebbins, pp. 147–72. New York: Academic Press; Baker, H. G. (1974). The evolution of weeds. *Annual Review of Ecology and Systematics*, 5, 1–24.

32. The most optimistic calculations come from Keeler, K. H. (1989). Can genetically engineered crops become weeds? *Bio/ Technology*, 7, 1134–9. Keeler suggested that the chance of turning a crop into a weed was of the order of 10^{-10}. This is a probability so low that it is difficult to imagine. In countries, such as the UK, that have a national lottery, it is hundreds of times less than the chance of your winning the top prize if you only buy one ticket in your lifetime.

33. Williamson, M. (1993). Reference 29.

34. Mooney, H. A. & Bernardi, G. (ed.) (1990). *Introduction of Genetically Modified Organisms into the Environment.* Chichester: John Wiley; Ginzburg, L. R. (ed.) (1991). *Assessing Ecological Risks of Biotechnology.* Boston: Butterworth-Heinemann; Wöhrmann, K. & Tomiuk, J. (ed.) (1993). *Transgenic Organisms: Risk Assessment of Deliberate Release.* Basel: Birkhäuser Verlag.

35. Crawley, M. J., Hails, R. S., Rees, M., Kohn, D. & Buxton, J. (1993). Ecology of transgenic oilseed rape in natural habitats. *Nature*, 363, 620–3.

36. Quoted by Watts, S. (1993). Organisms on the loose when red tape is cut. *Independent*, 25 October.

37. Miller, H. I. (1994). Risk-assessment experiments and the new biotechnology. *Trends in Biotechnology*, 12, 292–5.

38. Cavers, P. & Bough, M. (1985). Proso millet (*Panicum miliaceum* L.): a crop and a weed. In *Studies on Plant Demography: A Festschrift for John L. Harper*, ed. J. White, pp. 143–55. London: Academic Press.

39. Raybould, A. F. & Gray, A. J. (1993). Genetically modified crops and hybridization with wild relatives: a UK perspective. *Journal of Applied Ecology*, 30, 199–219.

40. de Vries, F. T., van der Meijden, R. & Brandenburg, W. A. (1992). Botanical files. *Gorteria*, suppl. 1, 1–100.

41. Dale, P. J., McPartlan, H. C., Parkinson, R., MacKay, G. R. & Scheffler, J. A. (1992). Gene dispersal from transgenic plants by pollen. In *The Biosafety Results of Field Tests of Genetically Modifed Plants and Microorganisms*, ed. R. Caspar & J. Landsmann, pp. 73–82. Braunschweig: Biologische Bundesanstalt für Land-Fortwirtschaft; Fishlock, D. (n.d.). *Environmentally Safe Crops for the Future: The PROSAMO Report – Testing the Environmental Impact of Plant Gene Technology*. Teddington, Middlesex: Laboratory of the Government Chemist.

42. Hoyle, R. (1994). Let's finally get the threat of virus-resistance plants straight. *Bio/Technology*, 12, 662–3.

43. Miller, H. I. (1993). Perception of biotechnology risks: the emotional dimension. *Bio/Technology*, 11, 1075–6.

44. Tepfer, M. (1993). Viral genes and transgenic plants: what are the potential environmental risks? *Bio/Technology*, 11, 125–32; Fox, J. L. (1994). USDA likely to okay Asgrow's engineered squash. *Bio/Technology*, 12, 761–2.

45. Anon (1993). New York City proposes first-of-kind law on posting of gentically engineered food. *geneWATCH*, 9(1–2), 9–11.

46. Jaffé, W. & Rojas, M. (1994). Transgenic potato tolerant to freezing. *Biotechnology and Development Monitor*, 18, 10. Note that this example is actually not one of genetic engineering of pest resistance in plants, rather one of the genetic engineering of yield enhancement by increasing the range of environments in which a plant can grow.

47. van Wijk. J. (1993). Farm seed saving in Europe under pressure. *Biotechnology and Development Monitor*, 17, 13–14.

48. Shiva, V. (1994). Freedom for seed. *Resurgence*, 163, 36–9.
49. Rural Advancement Foundation International (1994). *Conserving Indigenous Knowledge: Integrating Two Systems of Innovation*. New York: United National Development Programme.
50. Holmes, B. (1993). The perils of planting pesticides. *New Scientist*, 28 August, 34–7.
51. de Boef, W., Amanor, K., Wellard, K. & Bebbington, A. (1993). *Cultivating Knowledge: Genetic Diversity, Farmer Experimentation and Crop Research*. London: Intermediate Technology Publications; Holmes, B. (1994). Super rice extends limits to growth. *New Scientist*, 29 October, 4; Coghlan, A. (1994). Emergency rescue saves Poland's pine forest. *New Scientist*, 12 November, 26; Kidd, G. & Dvorak, J. (1994). A gutsy map of the future of agbiotech. *Bio/Technology*, 12, 1064–5.
52. Comstock, G. (1988). The case against bGH. *Agriculture and Human Values*, Summer, p. 39.
53. Taverne, D. (1990). *The Case for Biotechnology*, p. 43. London: Prima Europe.
54. Quoted in Juma, C. (1989). *The Gene Hunters*, p. 159. London: Zed Books.
55. Jones, I. H. (1991). European directive on patents: an introduction to the issues involved. Paper given to a Consultation on Moral and Theological Questions in Genetic Manipulation at Luton Industrial College, p. 21.
56. Rifkin, J. (1987). Biotechnology: major societal concerns. In *Public Perceptions of Biotechnology*, ed. L. R. Batra and W. Klassen, p. 55. Bethesda, MA: Maryland Agriculture Research Institute.
57. Jukes, T. H. (1988). Hazards of biotechnology: facts and fancy. *Journal of Chemical Technology and Biotechnology*, 43, 246.
58. Reference 53, p. 38.
59. Comstock, G. (1990). Reference 13, p. 128.
60. Comstock, G. (1990). Reference 13, p. 136.
61. Some of the material in this chapter has been drawn from Straughan, R. (1995). Ethical aspects of crop biotechnology.

In *Issues in Agricultural Bioethics*, ed. T. B. Mepham, G. A. Tucker & J. Wiseman. Nottingham, UK: Nottingham University Press.

7 THE GENETIC ENGINEERING OF ANIMALS

1. Whitelaw, B. (1995). Pharmaceuticals from transgenic sheep. *Biological Sciences Review*, 7(3), 25–7.
2. Kinsman, F. (1991). The recession. *Resurgence*, 147, 45.
3. British Medical Association (1992). *Our Genetic Future: The Science and Ethics of Genetic Technology*. Oxford: Oxford University Press.
4. James, R. (1993). Human therapeutic proteins generated in animals. *The Genetic Engineer and Biotechnologist*, 13, 189–97; Webster, J. (1993). Animal genetics – of pigs, oncomice and men. *Trends in Biotechnology*, 11, 1–2; Reference 1.
5. Cremers, H. C. (1993a). Transgenic bull to sow wild oats. *New Scientist*, 9 January, 8; Cremers, H. C. (1993b). Herman retains right to be a father. *New Scientist*, 8 May, 8.
6. Justice, M. J., Jenkins, N. A. & Copeland, N. G. (1992). Recombinant inbred mouse strains: models for disease study. *Trends in Biotechnology*, 10, 120–6; Porteous, D. J. & Dorin, J. R. (1993). How relevant are mouse models for human diseases to somatic gene therapy? *Trends in Biotechnology*, 11, 173–81; Anon (1995). Model mice. *New Scientist*, 11 February, 11.
7. Fox, J. L. (1993). Transgenic mice fall far short. *Bio/Technology*, 11, 663; Arthur, C. (1993). The onco-mouse that didn't roar. *New Scientist*, 26 June, 4.
8. Alper, J. (1993). Companies settle brave new frontier of xenografting. *Bio/Technology*, 11, 772–3.
9. Webb, J. (1993). Optimism over CF gene therapy. *New Scientist*, 20 March, 7. See also Lewin, R. (1992). Gene therapy promises cure for cystic fibrosis. *New Scientist*, 18 January, 9; Brown, P. (1992). Mutant mice offer hope for childhood disease. *New Scientist*, 19 September, 6.
10. Talk titled 'From pigs to Man: transplantation for the future' given by John Wallwork on 15 November 1993 to the

Cambridge branch for the Application of Research, Churchill
College, Cambridge; Talk titled 'Genetically engineering pigs
as a possible solution to the organ shortage crisis' given by
David White on 17 May 1995 to the Cambridge Medical
Society, Churchill College, Cambridge; White, D. &
Wallwork, J. (1993). Xenografting: probability, possibility, or
pipe dream? *Lancet*, 342, 879–80; Dunning, J., White, D. &
Wallwork, J. (1994). Transgenic pigs as potential donors for
xenografts. In *Rejection and Tolerance*, ed. J. L. Touraine *et al.*,
pp. 149–60. Netherlands: Kluwer Academic.

11. Concar, D. (1994). The organ factory of the future? *New
Scientist*, 18 June, 24–9.

12. Coghlan, A. (1993). Chickens lay golden eggs of genetics.
New Scientist, 20 November, 19; Sang, H. (1994). Transgenic
chickens – methods and potential applications. *Trends in
Biotechnology*, 12, 415–20; Love, J., Gribbin, C., Mather, C. &
Sang, H. (1994). Transgenic birds by DNA microinjection.
Bio/Technology, 12, 60–3.

13. Anon (1994). And the cow jumped over the moon. *GenEthics
News*, 3, 6–7; Maclean N. (ed.) (1994). *Animals with Novel
Genes*. Cambridge: Cambridge University Press.

14. Coghlan, A. (1993). Pressure group broods over altered
turkeys. *New Scientist*, 29 May, 9.

15. Dayton, L. (1992). 'Self-dipping' sheep will poison parasites.
New Scientist, 4 April, 19.

16. Reiss, M. J. (1993). Introductory overview: an ethical
framework for the use of animals in research and medicine. In
*Ethical Issues in Biomedical Sciences: Animals in Research and
Education*, ed. D. Anderson, M. Reiss & P. Campbell, pp. 3–
8. London: Institute of Biology.

17. Bentham, J. (1789). Principles of morals and legislation. In
The Collected Works of Jeremy Bentham, vol. 2.1, ed. J. H. Burns
& H. L. A. Hart, pp. 11–12. London: Athlone Press;
Bateson, P. (1991). Assessment of pain in animals. *Animal
Behaviour*, 42, 827–39; Smith, J. A. & Boyd, K. M. (ed.)
(1991). *Lives in the Balance: The Ethics of Using Animals in
Biomedical Research – The Report of a Working Party of the
Institute of Medical Ethics*. Oxford: Oxford University Press.

18. Dawkins, M. S. (1980). *Animal Suffering: The Science of Animal Welfare*. London: Chapman and Hall; Humphrey, N. (1986). *The Inner Eye*. London: Faber and Faber.
19. Hoyle, R. (1993). *Amicus* offers up disinformation and distortion. *Bio/Technology*, 11, 666–7 (particularly p. 666).
20. Ministry of Agriculture, Fisheries and Food (1995). *Report of the Committee to Consider the Ethical Implications of Emerging Technologies in the Breeding of Farm Animals*. London: HMSO.
21. Woychil, R. P., Stewart, T. A., Davis, L. G., D'Eustachio, P. & Leder, P. (1985). An inherited limb deformity created by insertional mutagenesis in a transgenic mouse. *Nature*, 318, 36–40; British Union for the Abolition of Vivisection & Compassion in World Farming (1993). *Oncomouse: Opposition Under Part V of the European Patent Convention*, Case Number T 19/90–3.3.2, Filing Number 85 304 490.7, Publication Number 0 169 672.
22. Linzey, A. (1987). *Christianity and the Rights of Animals*. London: SPCK; Linzey, A. (1994). *Animal Theology*. London: SCM; Animal Christian Concern, 46 St Margaret's Road, Horsforth, Leeds LS18 5BG, UK; Christian Consultative Council for the Welfare of Animals, 23 Ravensbourne Road, London SF8 4UU; Quaker Concern for Animal Welfare, Webb's Cottage, Saling, Braintree, CM7 5DZ, UK.
23. Folkert, K. W. (1991). Jainism. In *A Handbook of Living Religions*, ed. J. B. Hinnells, pp. 256–77. London: Penguin.
24. Midgley, M. (1983). *Animals and Why they Matter*, p. 61. Harmondsworth, UK: Penguin.
25. See arguments, on both sides of the case by Midgley (Reference 24); Regan, T. (1984). *The Case for Animal Rights*. Berkeley: University of California Press; Rachels, J. (1990). *Created from Animals: The Moral Implications of Darwinism*. Oxford: Oxford University Press; Barclay, O. R. (1992). Animal rights: a critique. *Science & Christian Belief*, 4, 49–61; Carruthers, P. (1992). *The Animals Issue: Moral Theory in Practice*. Cambridge: Cambridge University Press; Garner, R. (1993). *Animals, Politics and Morality*. Manchester: Manchester University Press; Linzey, A. (1994). *Animal Theology*. London: SCM.

26. Singer, P. (1975). *Animal Liberation: Towards an End to Man's Inhumanity to Animals*. (Paperback edition 1983). Wellingborough, UK: Thorsons.

27. For discussions of medical ethics see Kennedy, I. (1988). *Treat Me Right: Essays in Medical Law and Ethics*. Oxford: Clarendon Press; Beauchamp, T. L. & Childress, J. F. (1989). *Principles of Medical Ethics*, 3rd edn. Oxford: Oxford University Press; Shannon, T. A. (ed.) (1993). *Bioethics: Basic Writings on the Key Ethical Questions that Surround the Major, Modern Biological Possibilities and Problems*. Mahwah, NJ: Paulist Press.

28. Smith, J. A. & Boyd, K. M. (ed.) (1991). Reference 17, p. 323.

29. Spanner, D. C. (1992). [Book review of Rachels, J. (1990). *Created from Animals: The Moral Implications of Darwinism*. Oxford: Oxford University Press.] *Science & Christian Belief*, 4, 73–5. The term 'moral individualism' is introduced by Rachels in his book in a chapter titled 'Morality without the idea that humans are special'. Tay–Sachs disease is caused by a genetic mutation. It is particularly prevalent amongst Ashkenazi Jews. Individuals suffer progressive mental and motor deterioration and blindness. Fits often occur, there is no cure and the affected individual dies as a child.

30. Rifkin J., quoted in Reference 14.

31. Macer, D. (1990). Genetic engineering in 1990. *Science & Christian Belief*, 2, 25–40.

32. Mepham, T. B. (1994). Transgenesis in farm animals: ethical implications for public policy. *Politics and the Life Sciences*, 13, 195–203.

33. Holland, A. (1990). The biotic community: a philosophical critique of genetic engineering. In *The Bio-Revolution: Cornucopia or Pandora's Box?*, ed. P. Wheale & R. McNally, pp. 171–2. London: Pluto Press. See also Holland, A. (1995). Artificial lives: philosophical dimensions of farm animal biotechnology. In *Issues in Agricultural Bioethics*, ed. T. B. Mepham, G. A. Tucker & J. Wiseman, pp. 293–305. Nottingham: Nottingham University Press.

34. Verhoog, H. (1992). The concept of intrinsic value and

transgenic animals. *Journal of Agricultural and Environmental Ethics*, 5, 147–60; Verhoog, H. (1993). Morality and the 'naturalness' of transgenic animals. Paper presented at the Summer conference of the International Society for the History, Philosophy and Social Sciences of Biology, Boston, 13–18 July 1993.

35. Heideman, J. (1991). Transgenic rats: a discussion. In *Transgenic Animals*, ed. N. L. First & F. P. Haseltine, pp. 325–32 (see particularly p. 328). Boston: Butterworth-Heinemann.

36. Taylor, P. W. (1986). *Respect for Nature: A Theory of Environmental Ethics*. Princeton: Princeton University Press.

37. Reference 20, p. 1.

38. Reference 20, p. 15.

39. Ministry of Agriculture, Fisheries and Food (1993). *Report of the Committee on the Ethics of Genetic Modification and Food Use*. London: HMSO.

40. Submission dated 25 February 1993 from the Union of Muslim Organisations of U.K. & Eire to the Ministry of Agriculture, Fisheries and Food in connection with the Polkinghorne Committee.

41. Submission dated 23 January 1993 from Ranbir Singh Suri JP, Patron: Institute of Sikh Studies, to the Ministry of Agriculture, Fisheries and Food in connection with the Polkinghorne Committee.

42. Reference 39, p. 23.

43. See Sharpe, R. (1993). Danger – transplant scientist at work. *Outrage*, Jun/July, 6; Towell, J. (1993). Transgenic animal experiments. *Bio/Technology*, 11, 966; Kiernan, V. (1995). Artificial organs will trigger transplant boom. *New Scientist*, 21 January, 4.

44. Sharpe, R. (1993). Genetic nightmare for animals. *Outrage*, Jun/July, 12.

45. Shorrocks, B. & Coates, D. (ed.) (1993). *British Ecological Society-Ecological Issues No. 4: The Release of Genetically-engineered Organisms*. Montford Bridge, Shrewsbury: Field Studies Council; Maclean N. (ed.) (1994). Reference 13. See also Houck, M. A., Clark, J. B., Peterson, K. R. & Kidwell,

M. G. (1991). Possible horizontal transfer of *Drosophila* genes by the mite *Proctolaelaps regalis*. *Science*, 253, 1125–9.

46. For other attempts to formulate a set of ethical principles to govern the acceptability of the genetic engineering of animals, see the 1990 Report by the Advisory Committee in Ethics and Biotechnology in Animals, *Ethics and Biotechnology in Animals*. Wageningen, Netherlands: National Council for Agricultural Research (NRLO, the Netherlands); Mepham, T. B. (1994). Transgenesis in farm animals: ethical implications for public policy. *Politics and the Life Sciences*, 13, 195–203; and the two Christian responses to the Banner Committee available from The Arthur Rank Centre, National Agricultural Centre, Stoneleigh CV8 2LZ, UK and from The Church of Scotland Society, Religion and Technology Project, John Knox House, 45 High Street, Edinburgh EH1 1SR. For safety considerations see Berkowitz, D. B. & Kryspin-Sorensen, I. (1994). Transgenic fish: safe to eat? *Bio/Technology*, 12, 247–52.

8 THE GENETIC ENGINEERING OF HUMANS

1. Burnet, M. (1973). *Genes, Dreams and Realities*, p. 81. Harmondsworth, UK: Pelican. Sir Macfarlane Burnet was one of the most eminent geneticists of his time. He was awarded the Nobel Prize for medicine, held the Order of Merit and was created a KBE.
2. Postgate, J. (1995). Eugenics returns. *Biologist*, 45, 96.
3. Debenham, P. G. (1992). Probing identity: the changing face of DNA fingerprinting. *Trends in Biotechnology*, 10, 96–102; Inder, E. (1992). DNA fingerprinting: science, law, and the ultimate identifier. In *The Code of Codes: Scientific and Social Issues in the Human Genome Project*, ed. D. J. Kevles & L. Hood, pp. 191–210. Cambridge, MA: Harvard University Press.
4. Reiss, M. J. (1992). DNA Fingerprinting. In *Biology at Work*, ed. S. Tomkins, pp. 94–8. Cambridge: Cambridge University Press.
5. Lander, E. S. (1989). DNA fingerprinting on trial. *Nature*,

339, 501–5; Pringle, D. (1994). Who's the DNA
fingerprinting pointing at? *New Scientist*, 29 January,
51–2.
6. Glover, J. (1984). *What Sort of people Should There Be?*
Harmondsworth, UK: Penguin; Asimov, I. (1966). *Foundation
and Empire*. New York: Avon Books; Asimov, I. (1985). *Robots
and Empire*. New York: Granada Publishing; Leguine, U. K.
(1971). *City of Illusions*. London: Victor Gollancz; McCaffrey,
A. (1974). *To Ride Pegasus*. London: J. M. Dent. See also
McLean, S. A. M. (1994). *Law, ethics and the Human Genome
Project*. Society of Solicitors in the Supreme Courts of
Scotland Biennial Lecture. Available from Alistair R.
Brownlie, S.S.C. Library, 11 Parliament Square, Edinburgh
EH1 1RQ; McCrone, J. (1995). Watching you, watching me.
New Scientist, 20 May, 36–9.
7. Nuffield Council on Bioethics (1993). *Genetic Screening:
Ethical Issues*. London: Nuffield Council on Bioethics. See
also Clarke, A. (ed.) (1994). *Genetic Counselling: Practice and
Principles*. London: Routledge; Harris, J. (1992).
*Wonderwoman and Superman: The Ethics of Human
Biotechnology*. Oxford: Oxford University Press; Nelkin, D.
(1992). The social power of genetic information. In *The Code
of Codes: Scientific and Social Issues in the Human Genome
Project*, ed. D. J. Kevles & L. Hood, pp. 177–90. Cambridge,
MA: Harvard University Press.
8. Kiernan, V. (1995). US bans gene prejudice at work. *New
Scientist*, 2 April, 4.
9. Rawls, J. (1973). *A Theory of Justice*. Oxford: Oxford
University Press.
10. Reference 9, pp. 136–9.
11. For arguments in favour of the patenting of genes and
genetically engineered organisms, see Crespi, R. S. (1992).
What's immoral in patent law? *Trends in Biotechnology*, 10,
375–8; Scott-Ram, N. (1993). Making more of academic
assets. *Nature*, 364, 666–8; European Federation of
Biotechnology Task Group of Public Perceptions of
Biotechnology (1993). *Patenting Life*. London: National
Museum of Science and Industry; Pritchard, L. (1994).

From mice to medicines: protecting intellectual property. *Biologist*, 41, 149–52.

12. Brown, P. & Kleiner, K. (1994). Patent row splits breast cancer researchers. *New Scientist*, 24 September, 4.

13. For arguments against the patenting of genes and genetically engineered organisms see Patent Concern (n.d.). *Patenting of Plants and Animals*. London: The Genetics Forum; Bryant, J. (1992). Mapping the human genome: the Human Genome Project. *Science & Christian Belief*, 4, 105–25; Kimbrell, A. (1993). *The Human Body Shop: The Engineering and Marketing of Life*. London: HarperCollins*Religious*; Overseas Development Institute (1993). *Patenting Plants: The Implications for Developing Countries*. London: Overseas Development Institute; Montefiore, H. (1993). Rights in animals. *Church Times*, 19 February. For an account of the way Western researchers are rushing in to obtain, and sometimes patent, the genes of indigenous people, see Wilkie, T. (1995). *Gene Hunters*. London: Channel 4 Television; or contact the World Council of Indigenous Peoples, 2–100 Argyle Street, Ottawa K2P 1B6, Ontario, Canada.

14. See, for example, Burke, D. P. & McGough, K. J. (1993). PTO opens floodgates on animal patents. *Bio/Technology*, 11, 270; Coghlan, A. (1995). Europe kills off patents on life. *New Scientist*, 11 March, 7; Coghlan, A. (1995). Sweeping patent shocks gene therapists. *New Scientist*, 1 April, 4.

15. Morris, C. (1995). *MRC Research Update 3 – Cystic Fibrosis: The Quest for a Cure*. London: Medical Research Council.

16. Kohlberg, R. (1992). US alters cold virus to treat cystic fibrosis. *New Scientist*, 12 December, 5; Brown, P. (1993). Surprise illness halts gene therapy trial. *New Scientist*, 28 August, 5; Mitani, K. & Casakey, C. T. (1993). Delivering therapeutic genes – matching approach and application. *Trends in Biotechnology*, 11, 162–6; Anon (1995). *Lab Notes 2 – Gene Therapy: The Medical Use of Genes*. London: Wellcome Trust; Aldhous, P. (1995). Safer gene therapy in sight for cystic fibrosis. *New Scientist*, 7 January, 6. The May 1993 issue (volume 11, number 5) of *Trends in Biotechnology* is a special issue on gene therapy. Recent advances in the vectors

used to carry genes in gene therapy are reviewed by Hodgson, C. P. (1995). The vector void in gene therapy. *Bio/Technology*, 13, 222–5.

17. Watson, F. (1993). Human gene therapy – progress on all fronts. *Trends in Biotechnology*, 11, 114–7.
18. Edgington, S. M. (1993). Nuclease therapeutics in the clinic. *Bio/Technology*, 11, 580–2.
19. Anon (1995). Cystic fibrosis drug. *New Scientist*, 8 April, 11.
20. Fox, J. L. (1994). Gene therapy off to slow start. *Bio/Technology*, 12, 1066; Anon (1995). *Lab Notes 2 – Gene therapy: The Medical Use of Genes*. London: Wellcome Trust.
21. Meager, A. & Griffiths, E. (1994). Human somatic gene therapy. *Trends in Biotechnology*, 12, 108–13; Fox, J. L. (1993). Compassionate use of gene therapy stirs storm. *Bio/Technology*, 11, 252; Fox, J. L. (1993). NIHRAC approves 11 gene-therapy protocols. *Bio/Technology*, 11, 780; Menon, S. (1994). Herpes gene tackles brain tumours. *New Scientist*, 12 November, 11; Dodet, B. (1993). Commercial prospects for gene therapy – a company survey. *Trends in Biotechnology*, 11, 182–9.
22. Vines, G. (1994). Gene tests: the parents' dilemma. *New Scientist*, 12 November, 40–4; Emery, A. E. H. & Mueller, R. F. (1992). *Elements of Medical Genetics*, 8th edn. Edinburgh: Churchill Livingstone.
23. See, for example, Vines, G. (1993). Call to plan now for cancer gene screening. *New Scientist*, 1 February, 5; Miller, S. K. (1993). To catch a killer gene. *New Scientist*, 24 April, 37; Miller, S. K. (1993). Gene marker found for 'quiet scourge'. *New Scientist*, 15 May, 7; Miller, S. K. (1993). Alzheimer's gene 'the most important ever found'. *New Scientist*, 21 August, 17. In addition, newspapers frequently announce 'Gene for . . . found'.
24. Strohman, R. (1994). Epigenesis the missing beat in biotechnology. *Bio/Technology*, 12 February, 156–64. See also Horgan, J. (1993). Eugenics revisited. *Scientific American*, June, 92–100.
25. Brown, P. (1994). Breast cancer 'too complex' for gene test. *New Scientist*, 10 December, 4–5.

26. Miller, S. K. (1993). 'Colon gene' rounds off a brilliant year. *New Scientist*, 11 December, 4.
27. Miller, S. K. (1993). Search for genes that make a man's best friend. *New Scientist*, 26 June, 5.
28. See, for example, Charles, D. (1992). Genetics meeting halted amid racism charges. *New Scientist*, 26 September, 4; Kiernan, V. (1993). Panel hears conflicting views on biology of violence. *New Scientist*, 12 June, 7; Miller, S. K. (1993). Gene hunters sound warning over gay link. *New Scientist*, 24 July, 4–5; Mestel, R. (1994). The fearful gene in our children's make-up. *New Scientist*, 26 February, 10; Mestel, R. (1994). What triggers the violence within? *New Scientist*, 26 February, 31–4; Holmes, B. (1994). Gay gene test 'inaccurate and immoral'. *New Scientist*, 5 March, 9.
29. Charles, D. (1992). Reference 28.
30. This point is argued in detail by Rose, S., Lewontin, R. C. & Kamin, L. J. (1984). *Not in Our Genes: Biology, Ideology and Human Nature*. London: Penguin; and by Lewontin, R. C. (1993). *The Doctrine of DNA: Biology as Ideology*. London: Penguin. See also Sayers, J. (1992). *Biological Politics: Feminist and Anti-Feminist Perspectives*. London: Tavistock Publications; Plomin, R. (1994). *Genetics and Experience: The Interplay between Nature and Nurture*. Newbury Park, CA: Sage; Council for Responsible Genetics (1994). CRG platform statement: guiding the promise of biotechnology. *GeneWATCH*, 9(5–6), 5–9; Holmes, B. (1995). Why the search for smart genes is doomed. *New Scientist*, 4 March, 12.
31. Kennedy, I. (1988). *Treat Me Right: Essays in Medical Law and Ethics*. Oxford: Clarendon Press; Beauchamp, T. L. & Childress, J. F. (1989). *Principles of Medical Ethics*, 3rd edn. Oxford: Oxford University Press; Shannon, T. A. (ed.) (1993). *Bioethics: Basic Writings on the Key Ethical Questions that Surround the Major, Modern Biological Possibilities and Problems*. Mahwah, NJ: Paulist Press; Medical Research Council (1995). *Principles in the Assessment and Conduct of Medical Research and Publicising Results*. London: Medical Research Council.
32. Clothier Committee (1992). *Report of the Committee on the Ethics of Gene Therapy*. London: HMSO.

33. Reference 32, p. 17.
34. Brown, P. (1992). Britain dithers over gene therapy. *New Scientist*, 12 December, 4; Anon (1993). Britain embraces gene therapy. *New Scientist*, 27 February, 6.
35. Charles, D. (1992). Bush's 'ploy' on fetal tissue research. *New Scientist*, 30 May, 5; Anon (1993). Fetal funding. *New Scientist*, 30 January, 10.
36. Anon (1992). Transplant success. *New Scientist*, 5 December, 10.
37. *Horizon – Twice Born* shown on UK television, BBC2, 14 February 1995.
38. Coghlan, A. (1993). 'Gene gun' aims at Parkinson's. *New Scientist*, 24 April, 18.
39. Ferry, G. (1994). Parkinson's – a suitable case for treatment? *New Scientist*, 3 December, 36–40.
40. Holmes, J. & Lindley, R. (1991). *The Values of Psychotherapy*. Oxford: Oxford University Press.
41. Reference 32, p. 22.
42. Nelson, J. R. (1994). *On the New Frontiers of Genetics and Religion*, p. 151. Grand Rapids, MI: William B. Eerdmans.
43. In 1983, 58 scientific and religious leaders signed a statement calling for the banning of all germ-line research in humans (*Nature*, 309, 301–2 (1984)). See also World Council of Churches (1989). *Biotechnology: Its Challenges to the Churches and the World*. World Council of Churches, Subunit on Church and Society; Jochemsen, H. (1992). Medical genetics: its presuppositions, possibilities and problems. *Ethics & Medicine*, 8, 18–31.
44. For example, Suzuki, D. & Knudtson, P. (1988). *Genethics: The Ethics of Engineering Life*. Canada: Stoddart; British Medical Association (1992). *Our Genetic Future*. Oxford: Oxford University Press; 1994 Report of the Group of Advisors on the Ethical Implications of Biotechnology set up by Jacques Delors, the president of the European Commission. A number of feminist authors have been extremely suspicious about the genetic engineering of humans, suspecting that women are being used as guinea pigs or to produce 'perfect babies'. See Spallone, P. (1992).

Generation Games: Genetic Engineering and the Future for Our Lives. London: The Women's Press; Rowland, R. (1993). *Living Laboratories: Women and Reproductive Technology.* London: Cedar.

45. See, in particular, Glover, J., Reference 6; Harris, J., Reference 7. Warnock, M. (1992). Ethical challenges in embryo manipulation. *British Medical Journal*, 304, 1045–9.

46. Coghlan, A. (1994). Outrage greets patent on designer sperm. *New Scientist*, 9 April, 4–5.

47. This is not to imply that germ-line therapy will ever be totally risk free. As we discussed in Chapter 3, no activity, certainly no new technology, is ever 'totally safe'.

48. Barbour, I. G. (1992). *Ethics in an Age of Technology: The Gifford Lectures 1989–1991*, vol. 2, p. 197. London: SCM Press.

49. Nozick, R. (1974). *Anarchy, State, and Utopia.* New York: Basic Books.

50. Harris, J., Reference 7.

51. Harris, J. (1994). Biotechnology, friend or foe? Ethics and controls. In *Ethics and Biotechnology*, ed. A. Dyson & J. Harris, pp. 216–29. London: Routledge.

52. Harris, J., Reference 7, p. 178.

53. Reference 42, p. 164.

54. Text of the 1995 Heslington lecture given by John Habgood on 1 February at the University of York, pp. 10–11.

55. Glover, J., Reference 6, p. 149.

56. Aldhous, P. (1995). Fruit flies achieve total recall. *New Scientist*, 29 April, 18.

9 PUBLIC UNDERSTANDING OF GENETIC ENGINEERING: WHAT CAN EDUCATION DO?

1. Lawlor, E. (1989). *Individual Choice and Higher Growth*, p. 26. Luxembourg: Office for Official Publications of the European Comunities.

2. Healy, M. (1990). *Consumers and the Free Market*, p. 4. London: National Consumer Council.

3. See, for example, Hamstra, A. M. (1991). *Biotechnology in*

Foodstuffs, Dutch Institute for Consumer Research, SWOKA, which suggests that greater knowledge about genetic engineering does not necessarily lead to general acceptance *or* rejection, but rather to a more discriminating, differentiated approach which weighs each case on its perceived merits.

4. Macer, D. R. J. (1992). Public acceptance of human gene therapy and perceptions of human genetic manipulation. *Human Gene Therapy*, 3, 511–8; Lock, R. & Miles, C. (1993). Biotechnology and genetic engineering: students' knowledge and attitudes. *Journal of Biological Education*, 27, 267–72; Brown, C. M. (1994). *Consumer Attitudes to Biotechnology in Agriculture and Food: A Critical Review.* Letchmore Heath, Watford: IGD Business Publications; Rothenburg, L. (1994). Biotechnology's issue of public credibility. *Trends in Biotechnology*, 12, 435–8; Zechendorf, B. (1994). What the public thinks about biotechnology, *Bio/Technology*, 12, 870–5.

5. Straughan, R. (1989) *Beliefs, Behaviour and Education*, pp. 2–3. London: Cassell.

6. These distinctions are discussed in detail in, for example, Peters, R. S. (ed.) (1967). *The Concept of Education*. London: Routledge & Kegan Paul.

7. Schools Council Nuffield Humanities Project (1970). *The Humanities Project: An Introduction.* London: Heinemann.

8. See, for example, Taylor, M. (ed.) (1975). *Progress and Problems in Moral Education*, Part 2. Windsor, UK: NFER.

9. These presuppositions are well illustrated in John Wilson's detailed work on what constitutes a 'morally educated person'. See, for example, Wilson, J. (1990). *A New Introduction to Moral Education*, Part 3. London: Cassell.

10. These characteristics of ethical judgments and decisions are discussed in, for example, Straughan, R. (1988). *Can We Teach Children to be Good?*, Ch. 3. Milton Keynes: Open University Press.

11. See, in particular, Dearden, R. F. (1984). *Theory and Practice in Education.* London: Routledge; Bridges, D. (1986). Dealing with controversy in the curriculum: a philosophical perspective. In *Controversial Issues in the Curriculum*, ed. J. J. Wellington, pp. 19–38. Oxford: Basil Blackwell; Downie, R.

(1993). The teaching of bioethics in the higher education of biologists. *Journal of Biological Education*, 27, 34–8; Reiss, M. J. (1993). *Science Education for a Pluralist Society*. Buckingham: Open University Press.

12. Durant, J. (ed.) (1994). *UK Consensus Conference on Plant Biotechnology: Final Report*. London: Science Museum.

Index